| Centre number |
| Candidate number |
| Surname and initials |

Examining Group
General Certificate of Secondary Education

Physics
Higher Tier
Exam 1   Paper 1

Time: one and a half hours

Instructions to candidates

Write your name, centre number and candidate number in the boxes at the top of this page.

Answer ALL questions in the spaces provided on the question paper.

Show all stages in any calculations and state the units.
You may use a calculator.

Include diagrams in your answers where this may be helpful.

Information for candidates

The number of marks available is given in brackets **[2]** at the end of each question or part question.

The marks allocated and the spaces provided for your answers are a good indication of the length of answer required.

 Where you see this icon you will be awarded marks for the quality of written communication in your answers.
This means, for example, that you should:
- write in sentences
- use correct spelling, punctuation and grammar
- use correct scientific terms.

	For Examiner's use only
1	
2	
3	
4	
5	
6	
7	
8	
9	
10	
11	
Total	

© 2003 Letts Educational

EDUCATIONAL

Formula list

You may need to use this formula list to answer some questions.

power = current × voltage $\qquad\qquad\qquad\qquad P = I \times V$

voltage = current × resistance $\qquad\qquad\qquad V = I \times R$

average speed = distance travelled ÷ time taken $\quad v = \dfrac{s}{t}$

acceleration = increase in velocity ÷ time taken $\quad a = \dfrac{(v-u)}{t}$

pressure = force ÷ area $\qquad\qquad\qquad\qquad P = \dfrac{F}{A}$

work done or energy transfer = force × distance moved in its own direction
$$W = F \times d$$

charge = current × time $\qquad\qquad\qquad\qquad Q = I \times t$

$\dfrac{\text{primary voltage}}{\text{secondary voltage}} = \dfrac{\text{number of primary turns}}{\text{number of secondary turns}} \quad \dfrac{V_p}{V_s} = \dfrac{N_p}{N_s}$

force = mass × acceleration $\qquad\qquad\qquad F = m \times a$

wave speed = frequency × wavelength $\qquad\quad v = f \times \lambda$

Letts

1 A student set up a circuit which contained a motor.
The motor was used to work a model roundabout.
The student wanted to vary the speed of the motor with a variable resistor.

(a) Redraw the diagram to include the variable resistor and a voltmeter to record the voltage across the motor.

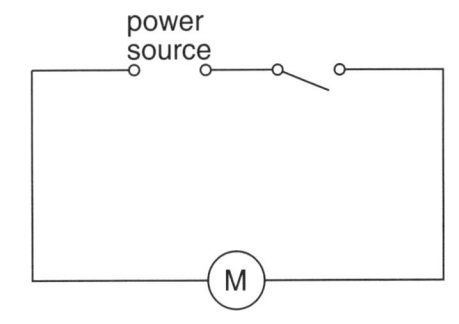

[3]

(b) Explain how the variable resistor controls the speed of the motor.

....

.... [2]

(Total 5 marks)

2 (a) The Earth attracts the moon with a gravitational force. The moon also attracts the Earth with a gravitational force.

Leave blank

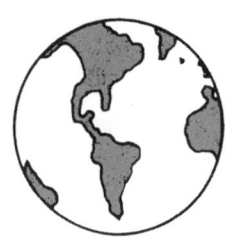

Write down two facts about this pair of forces:

1 The two forces are ....

2 The two forces are .... **[2]**

(b) An astronaut on earth has a mass of 90 kg.

(i) Using the equation:

weight = mass × gravitational field strength

calculate the weight of the astronaut on Earth
(gravitational field strength = 10 N/kg)

weight of astronaut on Earth = ... **[2]**

(ii) On the moon, what is the mass of the astronaut?

.... **[1]**

(iii) Gravitational field strength on the moon is 1.6 N/kg.
Calculate the weight of the astronaut on the moon.

weight of astronaut on the moon = ... **[2]**

(Total 7 marks)

© Letts Educational 2003

4

3 **(a)** Sound waves are longitudinal waves.
Light waves are transverse waves.
Describe the differences between longitudinal and transverse waves.

....

.... **[2]**

(b)

A candle held in front of a powerful loudspeaker flickers when loud music is played.
Explain, briefly, how this happens.

....

.... **[2]**

c) **(i)** In terms of frequency, what is the difference between audible sound and ultrasound?

.... **[1]**

(ii) An ultrasound transmitter and receiver on a trawler are used to detect a shoal of fish.

A pulse of ultrasound is detected 0.15 s after it is transmitted.
The speed of sound in water is 1500 m/s.
How far below the trawler is the shoal of fish?

.... **[3]**

(Total 8 marks)

4 (a) A student carried out an experiment using optical pins to find the position of an image in a plane mirror.

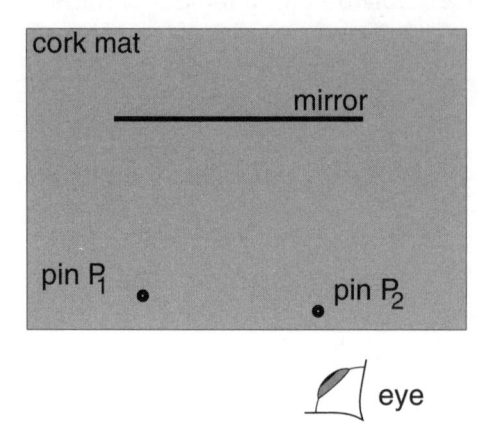

He looked into the mirror at the image of pin P_1 and placed pin P_2 exactly in line with the image of pin P_1.
Complete the diagram below to show:

1   the position of the image of pin P_1. Label it X.

2   an incident ray from pin P_1 and its reflected ray which arrives at pin P_2.

3   the normal at the point where the incident ray you have drawn meets the mirror.

_____

pin P_1 •   • pin P_2

[4]

(b) The image that the student observed is a **virtual** image.
What is meant by the term 'virtual'?

.... [1]

(c) Another student was carrying out reflection experiments with water waves in a ripple tank. For each experiment shown below, draw in **one** reflected wave to show the result she would obtain.

(i)

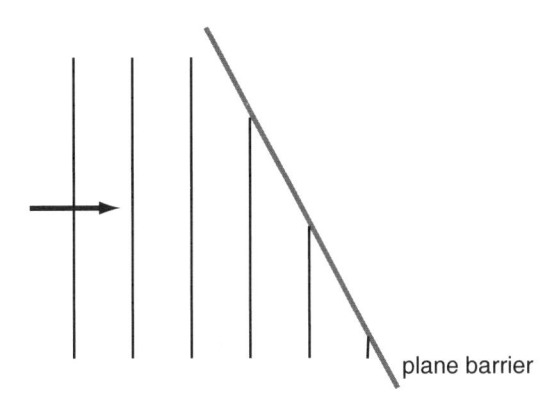

plane barrier

[1]

(ii)

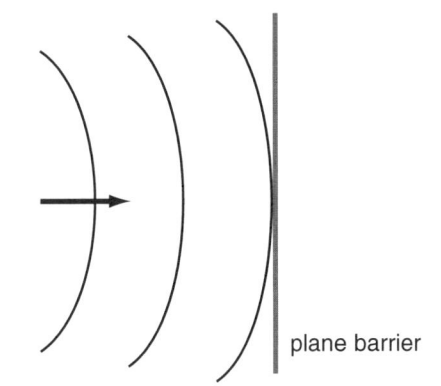

plane barrier

[1]

(iii)

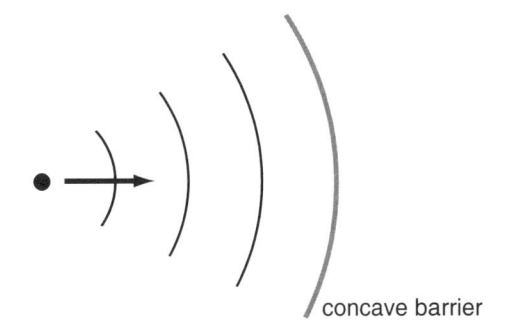

concave barrier

[1]

(Total 8 marks)

5 **(a)** A mug of cold tea can be reheated in a microwave oven. The tea becomes hot, but the mug handle can still be held comfortably. If you attempted to reheat the tea in an infrared oven, both the tea and the mug would be hot and the mug would be too hot to hold. Explain this difference in terms of the effects of microwaves and infrared.

Leave blank

....

....

.... **[3]**

(b) The wavelength of BBC Radio 4 is 1500 m. The speed of electromagnetic waves is 3×10^8 m/s. Calculate the frequency of the radio waves.

frequency = ... **[3]**

(c) Radio waves are suitable for broadcasting because they are readily diffracted.

(i) Complete the diagram to show diffraction (the diagram is a bird's eye view, looking down on the ground from above).

radio waves

barrier to radio waves (e.g. hills)

[2]

(ii) Explain why this diffraction effect makes radio waves suitable for broadcasting.

....

.... **[2]**

(Total 10 marks)

6 (a) There are three main types of radioactive emission, called alpha, beta and gamma. Alpha and beta are types of particle.

Leave blank

(i) Describe alpha and beta radiations in terms of atomic particles and charge:

alpha: type of particle ....

charge ....

beta: type of particle ....

charge .... [4]

(ii) Gamma radiation is not a stream of particles. What is gamma radiation?

.... [1]

(b) A radioactive source was placed in front of a Geiger tube which was connected to a counter. Different absorbers were placed, in turn, between the source and the Geiger tube. The following readings were taken:

absorber

source   Geiger tube

Absorber	Activity in counts/s
no absorber	540
brown paper	530
aluminium (5mm sheet)	80
lead (5mm sheet)	30

(i) Which type of radiation was being emitted from the source?

.... [1]

(ii) Explain your answer to part (i).

....

.... [2]

(iii) Explain how you would use the apparatus to measure background radiation.

....

.... [2]

(Total 10 marks)

7 (a) When a polythene strip is rubbed with a cloth, the polythene becomes negatively charged.

Explain, in terms of electrons, what happens to the polythene and the cloth.

....

....

....

.... **[3]**

(b) A nylon wig was placed on top of a metal dome. The metal was charged positively causing the wig to 'stand on end'.

metal dome

Explain what made the wig stand on end.

....

.... **[2]**

(Total 5 marks)

Letts

8 **(a)** A car was travelling at a constant velocity of 120 km/hour on a motorway.

  (i) What was the resultant force acting on the car?

  .... **[1]**

  (ii) The next day the same car did the same journey, at the same velocity, but used more fuel. All the conditions were the same except this time the car had two bicycles on its roof rack.

  Explain why the car used more fuel.
  Use your knowledge of force and energy.

  ....

  .... **[2]**

(b) The graph shows the motion of a cyclist going downhill.

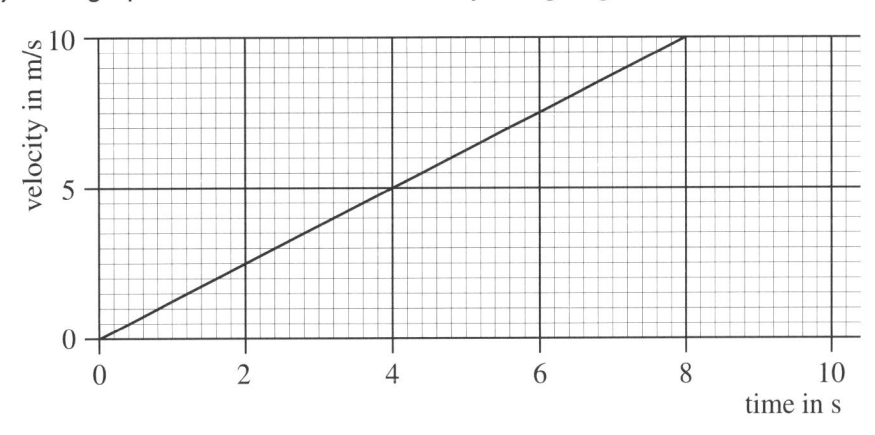

  (i) From the graph, calculate the acceleration of the cyclist. Show clearly on the graph how you obtained the necessary information.

  .... **[4]**

  (ii) The cyclist and his bicycle had a total mass of 900 kg.
  Calculate the force that produced the acceleration.

  force = ... **[3]**

  (Total 10 marks)

9 (a) Sketch the magnetic field patterns between the poles of the magnets shown below. Include arrows to show the directions of the fields.

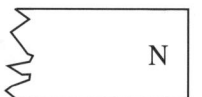 

[4]

(b) The magnetic field pattern around a current-carrying coil is similar to that of a bar magnet.

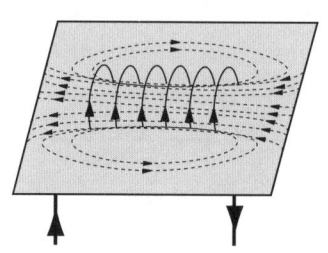

Suggest **three** ways in which the strength of the magnetic field can be increased.

1 ....

2 ....

3 .... **[3]**

Letts

c) A relay circuit is used to switch on a car starter motor.

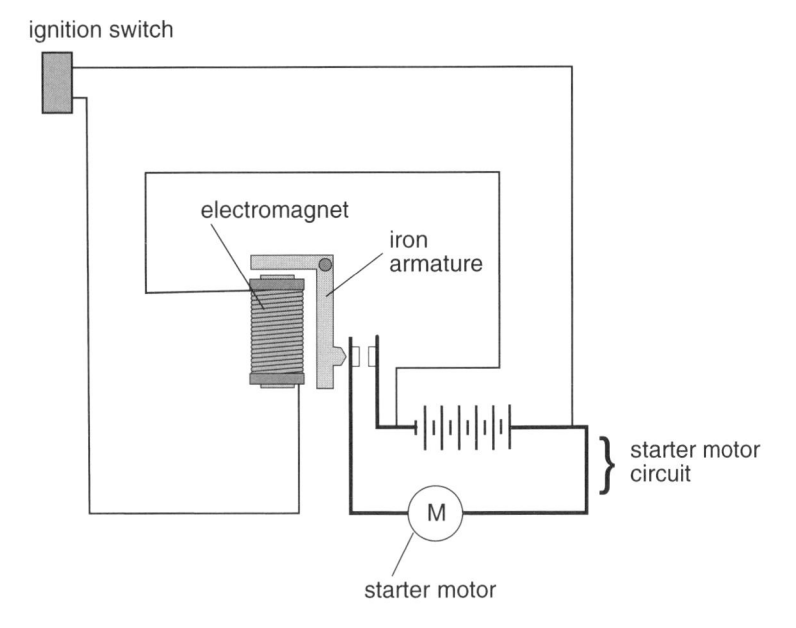

ignition switch

electromagnet

iron armature

starter motor circuit

M

starter motor

(i) Explain how this relay works when the ignition is switched on.

....

....

....

.... **[4+1]**

(ii) Why does the starter motor circuit have much thicker wires than the ignition switch circuit?

....

.... **[2]**

(Total 14 marks)

Letts

10 (a) In an experiment to stretch a helical spring by increasing the load on the spring, the following results were obtained:

load in N	length in cm	extension in cm
0.0	6.0	0.0
0.2	6.4	0.4
0.4	6.9	0.9
0.6	7.1	1.1
0.8	7.7	1.7
1.0	8.0	2.0
1.2	8.4	2.4

(i) Plot a graph of extension in cm against load in N.

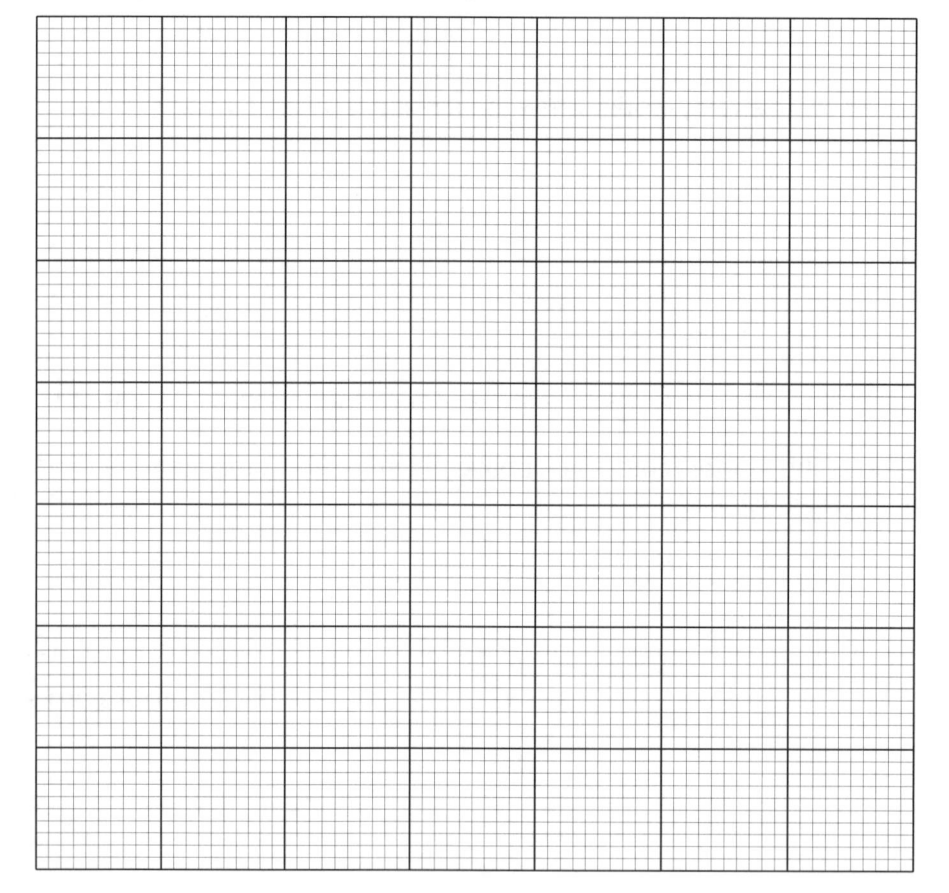

[4]

(ii) Write a conclusion about the behaviour of the spring.

.... [1]

(iii) From the graph, find the extension that would be produced by a load of 0.7 N. Show clearly on the graph how you obtained your answer.

....

.... [2]

(iv) To extend the spring by 2.4 cm an average force of 0.6 N is required. Calculate the work done by this force in stretching the spring. Write down the equation that you use.

equation ....

calculation

work done = ... **[3]**

(b) The Victorian game bagatelle has a spring which is compressed, then released to propel a marble.

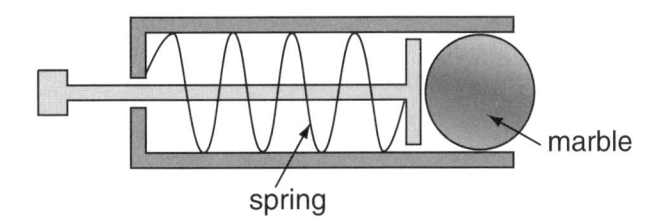

spring

marble

The work done in compressing the spring on a bagatelle board was 0.004 J. The marble had a mass of 0.004 kg. Assuming that all the energy stored in the spring is transferred to kinetic energy of the marble, calculate the velocity of the marble as it is released.

Use the equation:

$$\text{kinetic energy} = \frac{1}{2}mv^2$$

velocity of marble = ... **[3]**

(Total 13 marks)

11 (a) An electric heater has two heating elements.

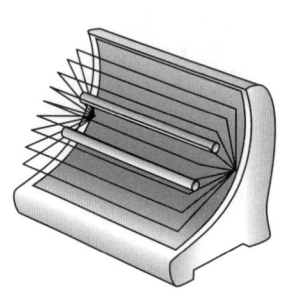

Explain why the elements are connected together in parallel.

.... **[1]**

(b) When one element is switched on, the voltage across it is 240 V and the current through it is 4 A. It is left on for 5 minutes.

(i) Calculate the power supplied to the element.

power = ... **[3]**

(ii) Calculate the total charge that flows through the element in the five minutes.

total charge = ... **[3]**

(iii) How many joules of energy are transferred for each coulomb of charge that passes between the ends of the element?

.... **[1]**

c) The heater is supplied with alternating current. Explain the difference between alternating current and direct current.

....

.... **[2]**

(Total 10 marks)

Centre number	
Candidate number	
Surname and initials	

 **Examining Group**

General Certificate of Secondary Education

Physics
Higher Tier
Exam 1   Paper 2

Time: one hour

Instructions to candidates

Write your name, centre number and candidate number in the boxes at the top of this page.

Answer ALL questions in the spaces provided on the question paper.

Show all stages in any calculations and state the units.
You may use a calculator.

Include diagrams in your answers where this may be helpful.

Information for candidates

The number of marks available is given in brackets **[2]** at the end of each question or part question.

The marks allocated and the spaces provided for your answers are a good indication of the length of answer required.

 Where you see this icon you will be awarded marks for the quality of written communication in your answers.
This means, for example, that you should:
- write in sentences
- use correct spelling, punctuation and grammar
- use correct scientific terms.

For Examiner's use only	
1	
2	
3	
4	
5	
6	
7	
Total	

 **EDUCATIONAL**

© 2003 Letts Educational

Formula list

You may need to use this formula list to answer some questions.

power = current × voltage $\qquad P = I \times V$

voltage = current × resistance $\qquad V = I \times R$

average speed = distance travelled ÷ time taken $\quad v = \dfrac{s}{t}$

acceleration = increase in velocity ÷ time taken $\quad a = \dfrac{(v-u)}{t}$

pressure = force ÷ area $\qquad P = \dfrac{F}{A}$

work done or energy transfer = force × distance moved in its own direction
$$W = F \times d$$

charge = current × time $\qquad Q = I \times t$

$\dfrac{\text{primary voltage}}{\text{secondary voltage}} = \dfrac{\text{number of primary turns}}{\text{number of secondary turns}} \qquad \dfrac{V_p}{V_s} = \dfrac{N_p}{N_s}$

force = mass × acceleration $\qquad F = m \times a$

wave speed = frequency × wavelength $\qquad v = f \times \lambda$

Letts

1 A popular toy in the past was the 'pop' gun. The diagram shows a 'pop' gun with a cork of mass 0.5 g which shoots out at a velocity of 3 m/s.

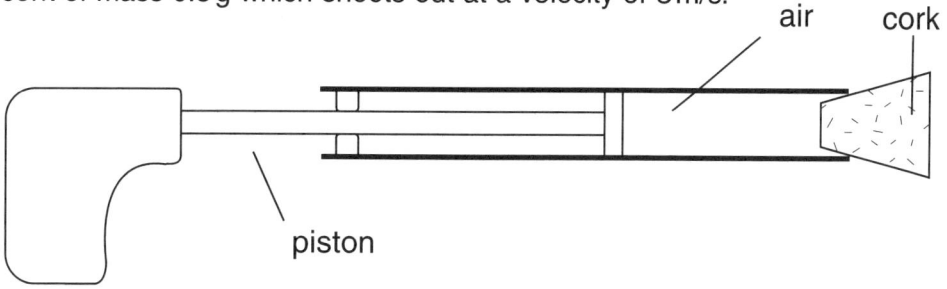

(a) The air in the tube was compressed until the pressure was high enough to shoot the cork out.

(i) What was the momentum of the cork before it left the gun?

.... **[1]**

(ii) Calculate the momentum of the cork immediately after it left the gun using the equation:

momentum = mass × velocity

momentum of cork = ... **[3]**

(iii) The answers for parts (i) and (ii) are not the same. Since momentum is conserved, what does this tell you about the movement of the gun as the cork shoots out?

....

.... **[2]**

(b) The volume of a fixed mass of gas at constant temperature is related to the pressure. The relationship is described by the equation:

pV = constant.

With the piston of the 'pop' gun pulled out as far as possible, the air in the tube is at a pressure of 1.1 Pa. The volume of the tube is 60 cm³. When the piston is pushed in halfway, so that the volume is 30 cm³ and the cork is still in position, what is the new pressure of the air in the tube?

pressure = ... **[3]**

(c) Use a particle model to explain the change in pressure when the volume is reduced.

....

....

.... **[3]**

(Total 12 marks)

2 A bridge was to be built across a river where there was road access from one side only. The bridge had to be built out from one side. The diagram shows how this was achieved.

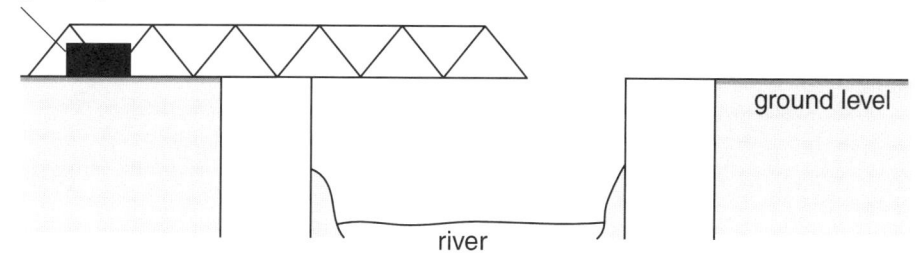

large weight

ground level

river

A laboratory model was made to test the idea. The model bridge weighed 50 N and was balanced on a pivot by using an additional weight **W**.

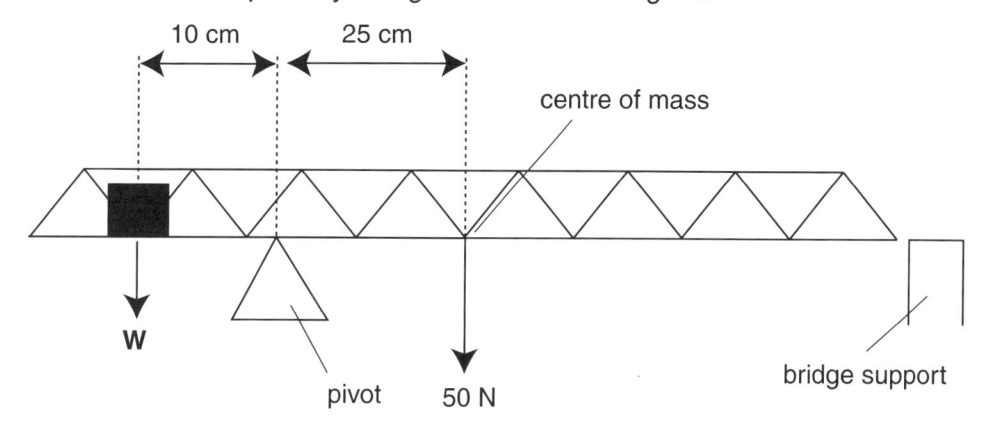

10 cm   25 cm

centre of mass

W

pivot   50 N

bridge support

(i) Explain the term 'centre of mass'.

....

.... **[1]**

(ii) Using the equation:
moment of a force = force × perpendicular distance to the pivot
calculate the moment of the weight of the bridge (50 N) about the pivot.

....

.... **[2]**

Letts

(iii) Using the equation:
clockwise moment = anticlockwise moment
calculate the value of the weight **W** required to balance the bridge model on the pivot.

....

....

.... **[3]**

(iv) When the bridge model is moved to the right (towards the bridge support), which way must the weight **W** be moved to maintain balance?

.... **[1]**

(Total 7 marks)

3 There are three types of radioactivity, α, β and γ, which have a variety of uses. The choice of source depends mainly on the half-life and the penetrating power required.

(a) What is meant by the term 'half-life'?

....

.... **[2]**

(b) The table below gives information about the half-lives of some radioactive sources.

SOURCE	TYPE OF RADIATION	HALF-LIFE
A	α	10 minutes
B	α	4 days
C	α	30 years
D	β	10 minutes
E	β	4 days
F	β	30 years
G	γ	10 minutes
H	γ	4 days
I	γ	30 years

State which source (**A** to **I**) you would choose for each use described below. Give reasons for your choice in terms of half-life and 'penetrating power'.

(i)

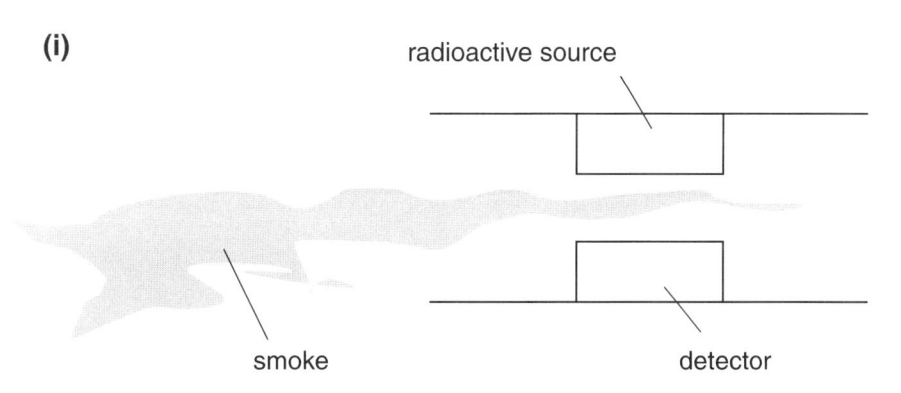

radioactive source

smoke   detector

A smoke detector: the alarm is set off when smoke prevents the radiation from the source reaching the detector which is a short distance (less than 5 cm) away.

choice of source: ....

reasons: ....

....

....

.... **[5+1]**

(ii)

detector

aluminium foil

radioactive source

A thickness gauge: the level of radiation reaching the detector monitors the thickness of the aluminium foil.

choice of source: ....

reasons: ....

....

....

.... **[5]**

(Total 13 marks)

4   The diagram shows a ray of light entering an optical fibre. The angle of incidence at point A is greater than the critical angle.

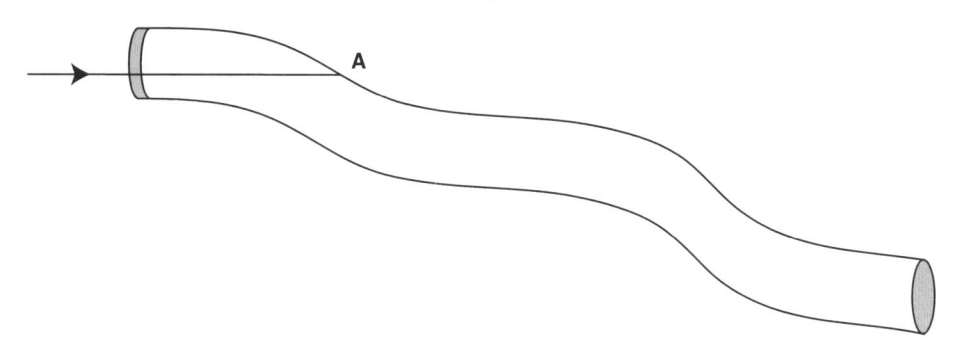

(a) (i)   On the diagram, show the path of the ray from point **A** until it emerges from the optical fibre.   [1]

   (ii)   What is the name given to the effect on the ray of light shown a point **A**?

....

.... [1]

(b) (i)   A student shone a ray of magenta light through a glass prism. The colour magenta is made up of blue and red light only. Complete the diagram to show the paths of the red and blue light through the prism and onto the screen. Label the rays **red** and **blue**.

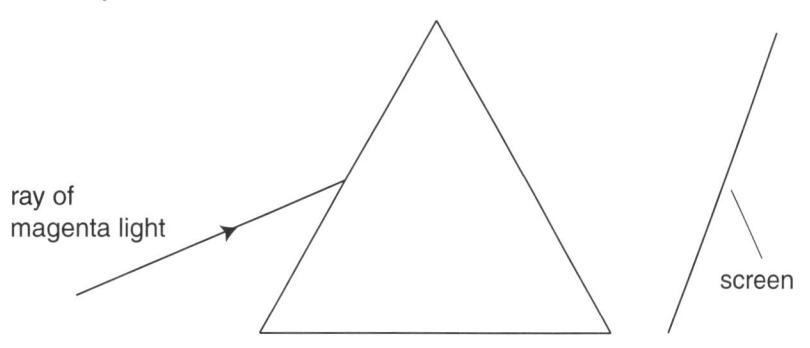

[3]

   (ii)   What is the name given to this effect of light being split into different colours?

....

.... [1]

(iii) Which of the following is a reason for the splitting of magenta light into red and blue as it passes through the prism? Tick your answer.

1   more red light than blue light is absorbed by glass ☐

2   red and blue light travel at different speeds through glass ☐

3   the red and blue light have different intensities ☐ **[1]**

(Total 7 marks)

5 **(a)** What is the difference between speed and velocity?

Leave blank

....

.... [2]

(b) **(i)**

A cyclist changes speed from 3 m/s to 5 m/s over a period of 12 s. Calculate the acceleration.

acceleration = ... [3]

(ii) The cyclist had an instrument on his bicycle which displayed his acceleration. At one point the display read ' −0.85'.

What does the negative sign tell you about the motion of the bicycle?

....

.... [2]

(Total 7 marks)

6 A student carried out an experiment to study the relationship between voltage and current of a filament lamp, using the circuit shown.

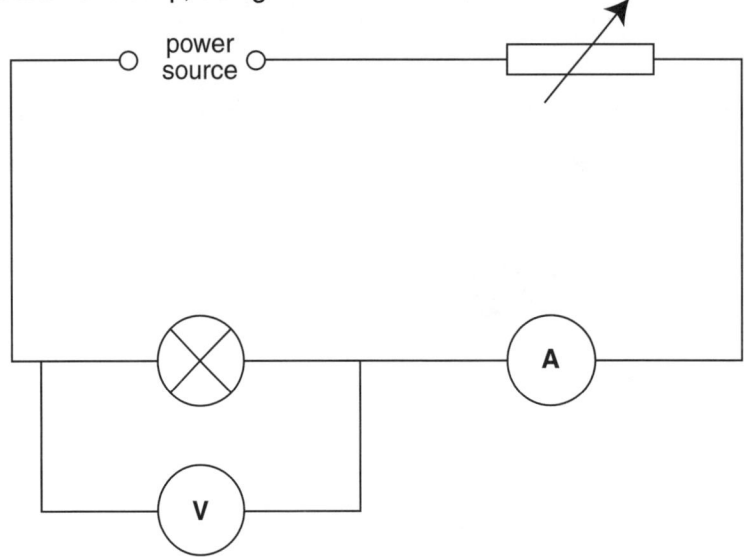

The readings obtained were as follows:

voltage in V	current in A
0	0
2	0.55
4	1.04
6	1.36
8	1.64
10	1.80

(i) Plot the graph of voltage against current. Draw the best fit curve.

[4]

Letts

(ii) As the current through the lamp increases, the filament temperature increases. What effect does this have on the resistance of the filament?

.... **[1]**

(iii) How could you tell that the filament temperature was increasing without using any measuring instruments?

.... **[1]**

(iv) On the diagram draw in the position of the pointer on the ammeter when the current reading is 1.42 A.

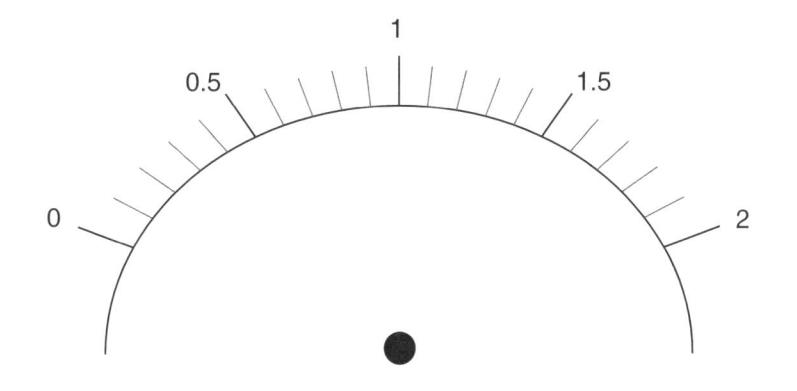

[1]

(Total 7 marks)

The diagram below shows the arrangement of particles in a solid.

Leave blank

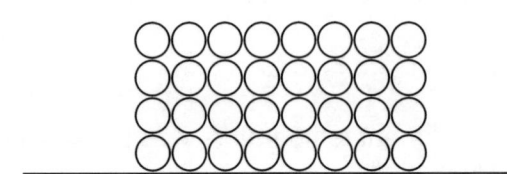

Complete diagrams (i) and (ii) to show the particles in a liquid and a gas.

(i) 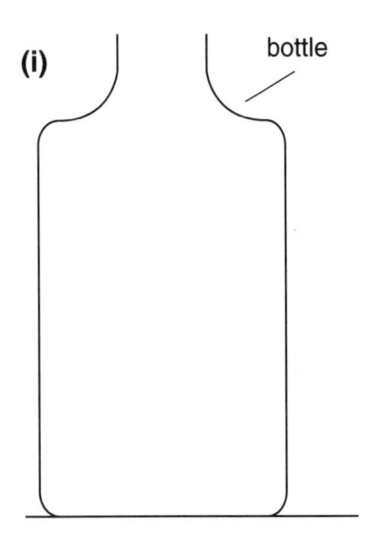 bottle

(ii) 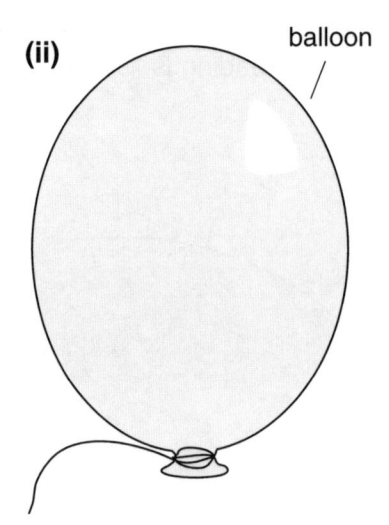 balloon

[4]

(iii) Describe, briefly, the movement of the particles in solids, liquids and gases.

solids: ....

....

liquids: ....

....

gases: ....

.... [3]

(Total 7 marks)

Letts

BLANK PAGE

© Letts Educational 2003

[turn over

BLANK PAGE

| Centre number |
| Candidate number |
| Surname and initials |

 **Examining Group**

General Certificate of Secondary Education

Physics
Higher Tier
Exam 2 Paper 1

Time: one and a half hours

Instructions to candidates

Write your name, centre number and candidate number in the boxes at the top of this page.

Answer ALL questions in the spaces provided on the question paper.

Show all stages in any calculations and state the units.
You may use a calculator.

Include diagrams in your answers where this may be helpful.

Information for candidates

The number of marks available is given in brackets **[2]** at the end of each question or part question.

The marks allocated and the spaces provided for your answers are a good indication of the length of answer required.

 Where you see this icon you will be awarded marks for the quality of written communication in your answers.
This means, for example, that you should:
- write in sentences
- use correct spelling, punctuation and grammar
- use correct scientific terms.

For Examiner's use only	
1	
2	
3	
4	
5	
6	
7	
8	
Total	

EDUCATIONAL

© 2003 Letts Educational

1 (a) The diagram below shows a wind farm using several wind turbines to generate electricity.

(i) What change of energy takes place in a wind turbine?

.... **[1]**

(ii) Name one disadvantage of using wind power.

.... **[1]**

(iii) Is wave energy a renewable or non-renewable source?

.... **[1]**

(iv) Each wind turbine is 74% efficient and generates 50 kW of electrical power.

Determine the input power to each wind turbine.

input power = ... kW **[3]**

Letts

(b) The diagram below shows a fairground ride.

Leave blank

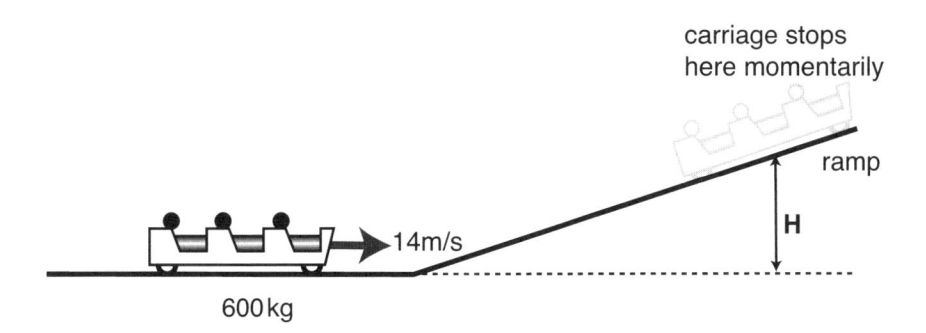

The total mass of the carriage and its occupants is 600 kg.
The velocity of the carriage at the bottom of the ramp is 14 m/s.
The carriage climbs up a ramp and momentarily stops at a vertical height H.

(i) Calculate the kinetic energy of the carriage and its occupants at the bottom of the ramp.

kinetic energy = ………… unit: …………**[4]**

(ii) The carriage comes to rest at the top of the ramp. What is the gain in gravitational potential energy of the carriage and its occupants? Assume there are no losses due to friction.

.... **[1]**

(iii) Calculate the height H of the carriage when it comes to rest. The gravitational field strength g = 10 N/kg.

H = ………… m **[3]**

(Total 14 marks)

© Letts Educational 2003   3   **[turn over**

2 The diagram below shows a 62 kg swimmer standing still on the edge of a diving board.

Leave blank

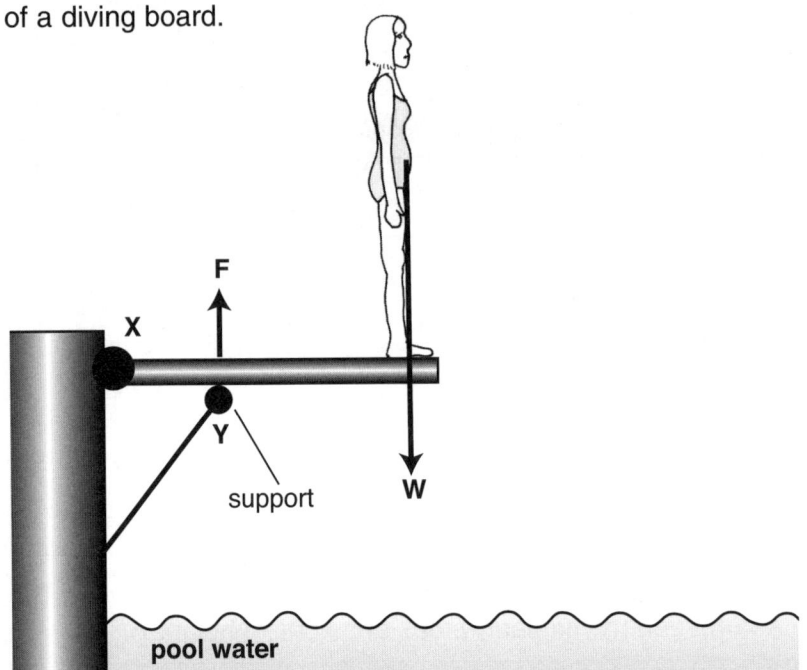

One of the forces acting on the swimmer is her weight. What other force acts on the swimmer as she stands still at the edge of the diving board?

.... **[1]**

(b) Calculate the weight W of the swimmer.
The gravitational field strength $g = 10\,\text{N/kg}$.

W = ... N **[2]**

(c) (i) The weight W of the swimmer exerts a moment about the point **X**.
State whether this moment is clockwise or anticlockwise.

.... **[1]**

(ii) The diving board is resting on the support **Y**. Discuss whether the force F at this support is greater than or less than the weight of the swimmer.

....

....

.... **[2]**

(d) The swimmer gently walks over the edge of the diving board and drops vertically down into the pool water below. The velocity against time graph below shows the motion of the swimmer in free fall and in the water.

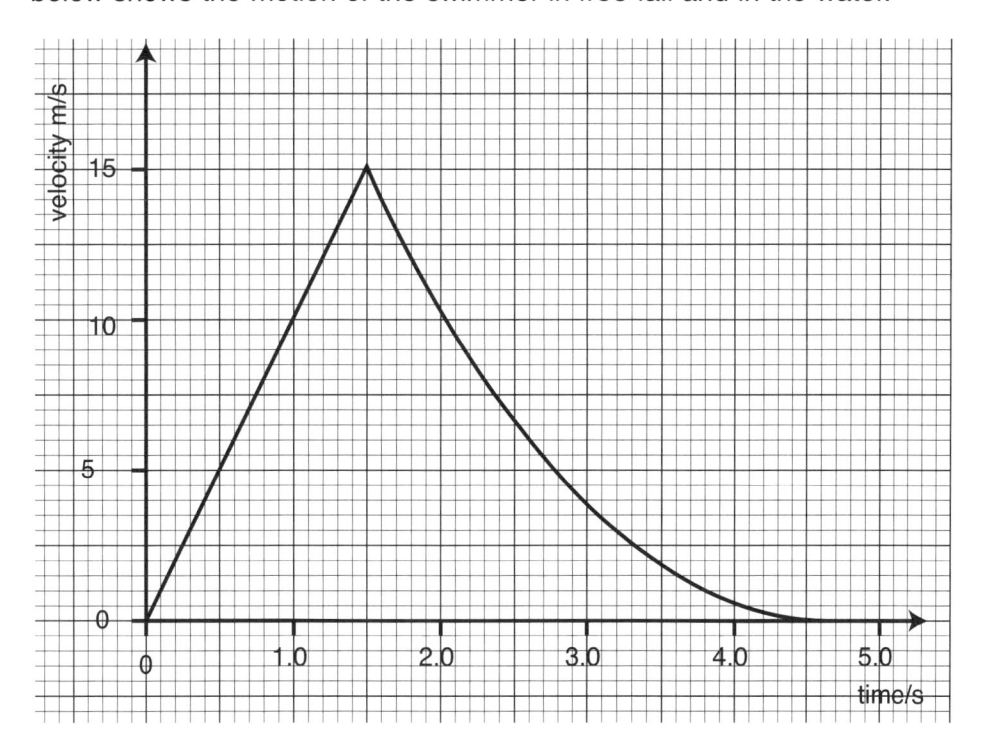

Use the graph

(i) to describe the motion of the swimmer,

....

....

....

.... [4+1]

(ii) to calculate the initial acceleration of free fall of the swimmer.

acceleration = ………….. m/s² **[3]**

(Total 14 marks)

3 (a) Visible light is a transverse wave. Explain what is meant by a transverse wave.

Leave blank

....

.... [1]

(b) The diagram below shows part of a rear reflector of a car.

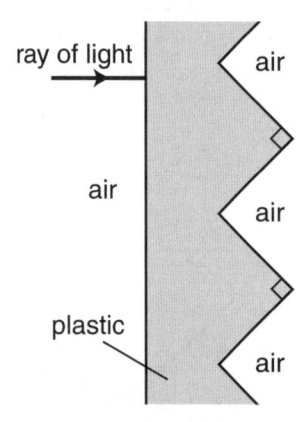

The reflector is made from red coloured plastic that is shaped as shown in the diagram so that the light from the rear is totally internally reflected.

(i) What happens to the speed of light when it travels from the air into the plastic?

.... [1]

(ii) Complete the path of the ray of light in the diagram above. [2]

(c) The diagram below shows a piano being played in a room where the door has been left open.

Suggest why a person on the other side of the open door can hear the piano clearly even though he cannot directly see it.

Leave blank

....

....

.... **[3]**

(d) The diagram below shows two loudspeakers connected to the same a.c. supply.

a.c. supply
220 Hz

A

B

loudspeakers

X

person listening
to sound from
loudspeakers

(i) With the switch **A** closed, and switch **B** open, a person at point **X** hears a very loud sound. The frequency of sound is 220 Hz. Calculate the wavelength of the sound given the speed of sound in air is 340 m/s.

wavelength = ………….. m **[3]**

(ii) When both switches **A** and **B** are closed, the person suddenly hears no sound even though each loudspeaker is still emitting sound. Explain why this happens.

....

....

.... **[2]**

(Total 12 marks)

4 (a) Complete the ray diagram below to show what is meant by the focal length of a diverging (concave) lens. **[2]**

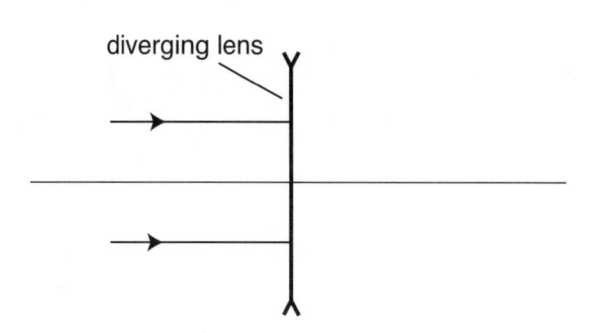

diverging lens

(b) The diagram below shows a converging (convex) lens used as a magnifying glass.

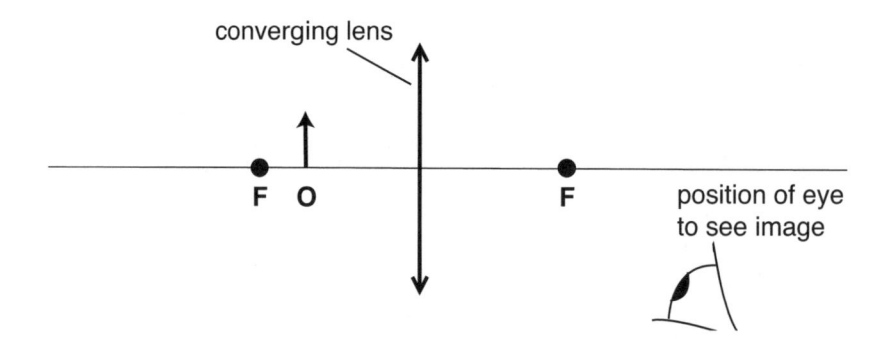

converging lens

F O

F

position of eye to see image

The point marked **F** is the principal focus of the converging lens.
The object **O** lies between the principal focus and the centre of the lens.
The image produced by the lens is virtual.

(i) Explain what is meant by a *virtual image*.

....

.... **[1]**

(ii) Draw a ray diagram to locate the position of the virtual image formed by the lens. Label the image **I**. **[3]**

(iii) Apart from being a virtual image of the object **O**, state two other properties of this image.

....

....

.... **[2]**

(Total 8 marks)

Letts

5 (a) The diagram below shows some sausages being heated under the heating element of a grill.

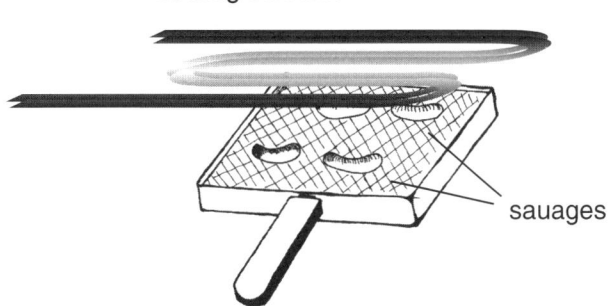

heating element

sauages

The heating element is at a temperature of about 1500 °C and the surface of the sausages is at about 220 °C. The electrical power rating of the heating element is 1.2 kW. The sausages take 10 minutes to cook.

(i) Name the process by which most of the heat reaches the sausages.

.... **[1]**

(ii) Name the process by which the inside of the sausages become hot.

.... **[1]**

(iii) Calculate the cost of cooking the sausages.
Each Unit of electricity (kWh) costs 7.2 pence.

cost = ... pence **[3]**

Letts

(b) The diagram shows two resistors connected to a 12 V battery.

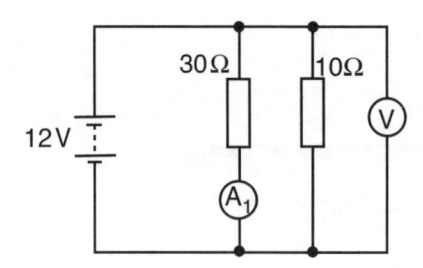

(i) Are the resistors connected in series or in parallel?

.... **[1]**

(ii) Calculate the current measured by the ammeter A_1.

current = ... A**[3]**

(iii) Calculate the heat energy released by the 30 Ω resistor in a time of 1 minute.

energy = ... J**[3]**

(Total 12 marks)

6 **(a)** The diagram below shows two coils made from insulated copper wires wrapped round a soft-iron rod.

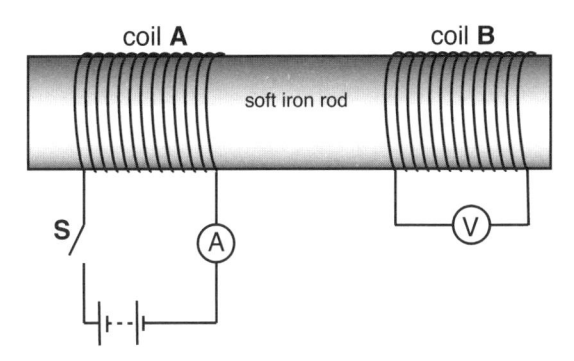

The coil **A** is connected in series with a switch **S**, an ammeter and a battery. The coil **B** is connected to a voltmeter.

The switch **S** is closed. The voltmeter connected to coil **B** shows a deflection and after a short period of time it shows no reading even though there is constant current shown by the ammeter. Explain these observations.

....

....

....

....

....

....

....

....

.... [4+1]

Letts

(b) A mobile phone charger unit has a transformer that steps-down the voltage from 230 V to 3.8 V. The primary coil has 5200 turns and an input current of 12 mA. The transformer is 100% efficient.

Calculate

(i) the number of turns on the secondary coil,

turns = ... [3]

(ii) the current in the secondary coil.

current = ... A [3]

(c) Electric power is distributed by the National Grid at a very high voltage of 400 000 V. Suggest why this voltage is so high.

....

.... **[1]**

(Total 12 marks)

7 Most of the carbon dioxide in the Earth's atmosphere contains atoms of the stable isotope carbon-12. A small percentage of the carbon dioxide also contains radioactive atoms of the isotope carbon-14.

(a) Explain what is meant by an *isotope*.

....

.... **[2]**

(b) Explain what is meant by background radiation. Name one source for this radiation.

....

.... **[2]**

(c) Carbon dioxide is absorbed by all living trees. The ratio of carbon-12 atoms to carbon-14 atoms in all living trees is a constant. When a tree dies it stops absorbing carbon-14 from the atmosphere. The carbon-14 already in the tree starts to decay. The half-life of carbon-14 atoms is 5600 years.

(i) An atom of carbon-14 may be represented as $^{14}_{6}$ C. In the nucleus of carbon-14, how many protons and neutrons are there?

protons: ....

neutrons:........... **[2]**

(ii) An archaeologist discovers a wooden spear. A sample of 1.0 gram of carbon taken from the wooden spear gives an average of 22.5 counts per hour and a 1.0 gram sample of carbon taken from a living tree gives an average of 90 counts per hour. Assuming that the ratio of carbon-12 to carbon-14 atoms has remained constant since the spear was made, determine the age of the wooden spear in years.

age = ... years **[3]**

(Total 9 marks)

© Letts Educational 2003

[turn over

8 **(a)** Astronomers believe that the Sun and the planets were formed from dust particles and atoms, which were attracted together.

Leave blank

(i) State one major difference between the Sun and the planets in our solar system.

.... **[1]**

(ii) Name the force responsible this attraction.

.... **[1]**

(b) Explain what is meant by an artificial satellite and suggest one of its uses.

....

....

.... **[2]**

(c) The diagram below shows an artificial satellite orbiting the Earth in a circular orbit of radius 6.7×10^6 m.

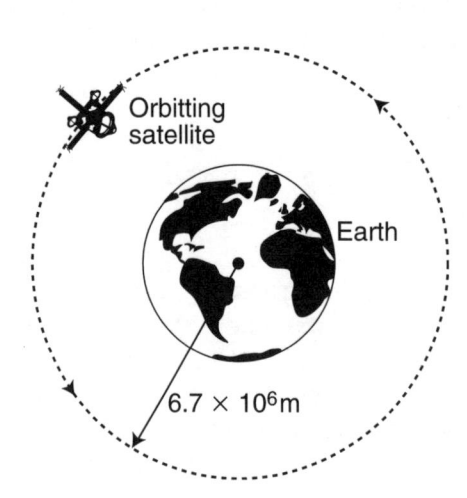

Orbitting satellite

Earth

6.7×10^6 m

The speed of the satellite is constant at 7.7 km/s.

(i) Explain why the velocity of the satellite changes even though its speed remains constant.

....

.... **[1]**

Letts

(ii) The centripetal acceleration a of the satellite is related to its orbital speed v and radius of orbit r by the equation

$$a = \frac{v^2}{r}$$

   1 Explain what is meant by *centripetal acceleration*.

.... **[1]**

   2 Determine the centripetal acceleration a of the satellite.

a = ... m/s^2 **[3]**

(Total 9 marks)

BLANK PAGE

| Centre number |
| Candidate number |
| Surname and initials |

 **Examining Group**

General Certificate of Secondary Education

Physics
Higher Tier
Exam 2   Paper 2

Time: one and a half hours

Instructions to candidates

Write your name, centre number and candidate number in the boxes at the top of this page.

Answer ALL questions in the spaces provided on the question paper.

Show all stages in any calculations and state the units.

You may use a calculator.

Include diagrams in your answers where this may be helpful.

Information to candidates

The number of marks available is given in brackets **[2]** at the end of each question or part question.

The marks allocated and the spaces provided for your answers are a good indication of the length of answer required.

 Where you see this icon you will be awarded marks for the quality of written communication in your answers.

This means, for example, that you should:

- write in sentences
- use correct spelling, punctuation and grammar
- use correct scientific terms.

For Examiner's use only	
1	
2	
3	
4	
5	
6	
7	
Total	

© 2003 Letts Educational

EDUCATIONAL

1 (a) The diagram below shows a child on a park swing.

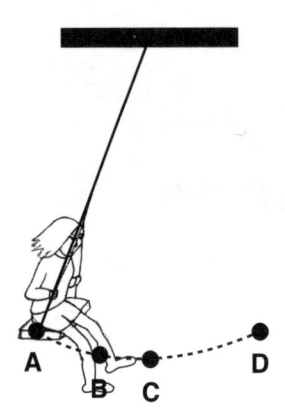

(i) On the diagram above, state in which position **A**, **B**, **C** or **D** would the air resistance on the child be a maximum.

.... **[1]**

(ii) Discuss the energy changes taking place as the child oscillates freely and comes to rest after some time.

....

....

....

.... **[2+1]**

(iii) The rubber seat of the swing hangs from two sets of metal chains. Explain why the metal chains feel cold when the child grips them with both hands, but the rubber seat does not.

....

....

.... **[2]**

(b) The diagram below shows the main losses of thermal energy from a house.

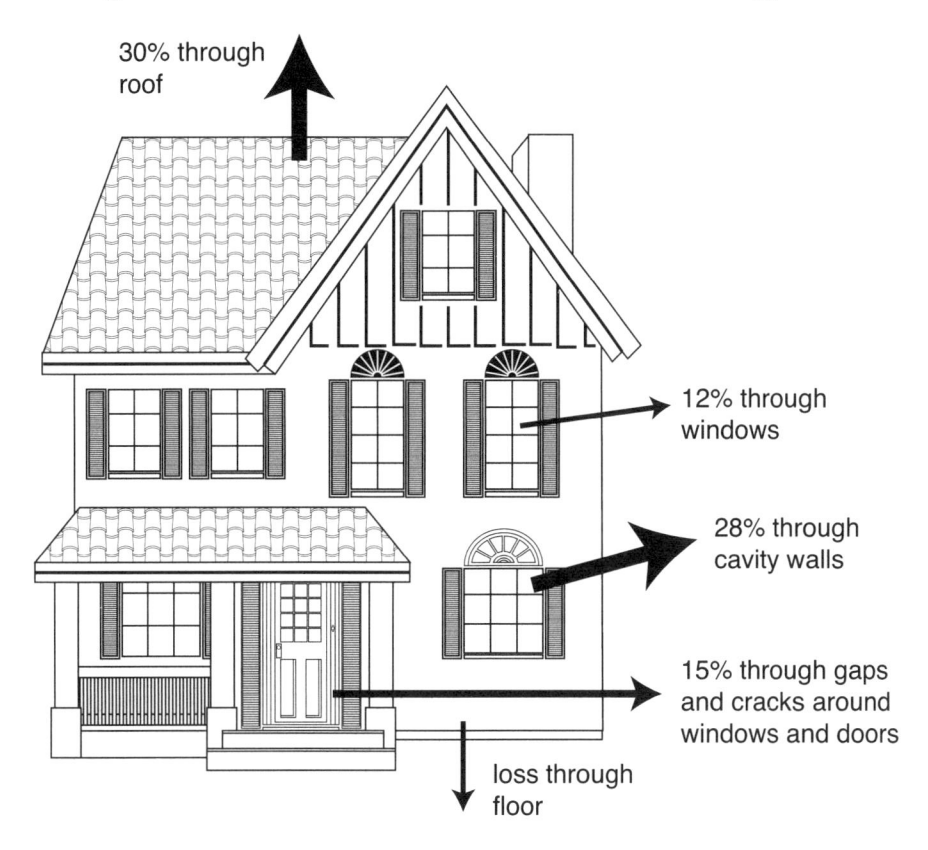

30% through roof

12% through windows

28% through cavity walls

15% through gaps and cracks around windows and doors

loss through floor

(i) Determine the percentage of heat loss from the floor of the house.

loss = ... % **[1]**

(ii) How can the heat loss through the flooring be reduced?

.... **[1]**

(iii) The windows in the house are double-glazed.
The diagram below shows a section through a double-glazed window.

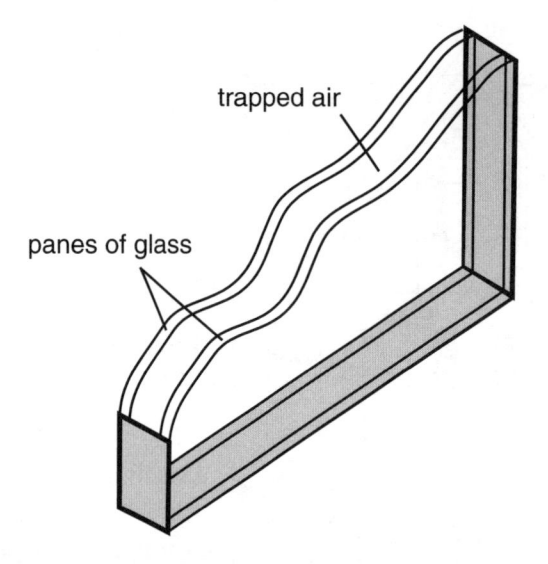

Discuss how the heat loss through the double-glazed window is reduced compared to a window that has a single pane of glass.

....

....

.... **[2]**

(iv) Explain why fitting shiny aluminium foil behind the radiators in the house would reduce heat loss.

....

.... **[1]**

(Total 11 marks)

2 (a) (i) Explain what is meant by the terms *thinking distance* and *braking distance* when describing the stopping distance of a car.

....

....

.... **[2]**

(ii) Discuss one factor that affects the braking distance of a car.

....

....

.... **[2]**

(b) A 940 kg car is travelling on a level road at a constant velocity of 20 m/s. The driver brakes suddenly and comes to rest in a time of 4.2 s after the brakes are applied.

(i) Calculate the deceleration of the car.

deceleration = ...m/s² **[3]**

(ii) Calculate the braking force acting on the car.

force = ... N **[3]**

The velocity against time graph for the braking car is shown below.

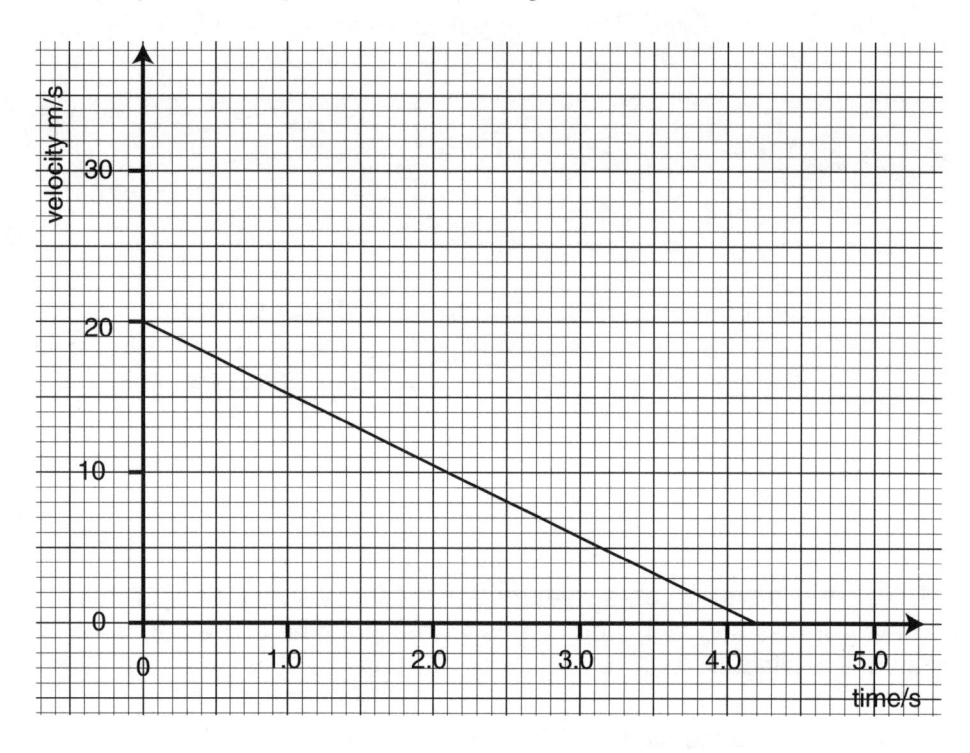

(iii) Use the graph to determine the braking distance of the car.

distance = ………………………….. m **[3]**

(iv) Calculate the work done by the braking force.

work done = …………… unit: ……… **[4]**

(v) Without any further calculations, suggest how the braking distance will change if the car was travelling at an initial speed of 40 m/s. You may assume that the braking force on the car remains constant.

…………………………………………………………………………………

…………………………………………………………………………………

…………………………………………………………………………………… **[2]**

(Total 19 marks)

Letts

3 **(a)** Suggest why sound waves cannot travel from the Earth to the moon.

....

.... **[1]**

(b) Explain what is meant by ultrasound.

....

.... **[1]**

(c) The diagram below shows an ultrasound scanner used in the aeronautics industry to detect cracks or flaws within the metal wings.

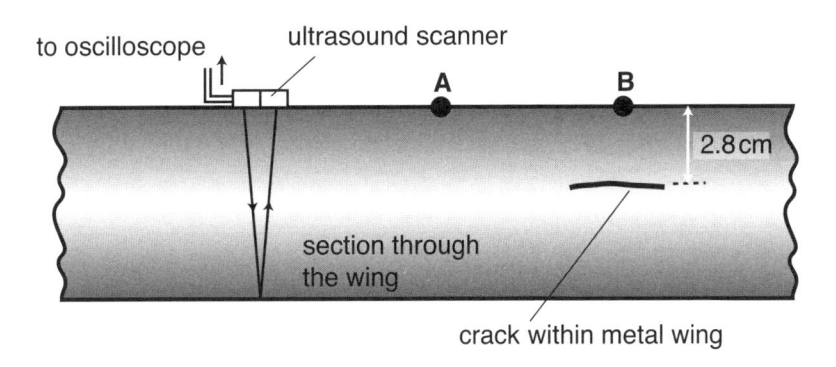

The scanner consists of an ultrasound transmitter and a receiver. The signals are displayed on an oscilloscope.

(i) The transmitter is placed at position **A**. The oscilloscope trace below shows the transmitted pulse **T** and the reflected pulse **R** from the bottom surface of the metal sample.

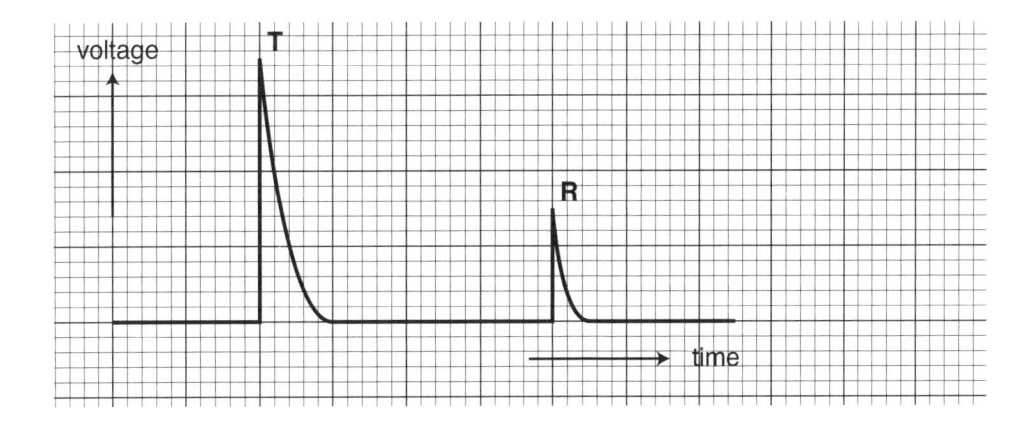

Suggest why the reflected signal has smaller amplitude.

....

.... **[1]**

(ii) The transmitter is now placed at position **B**. The oscilloscope trace below shows the transmitted pulse and the received pulses.

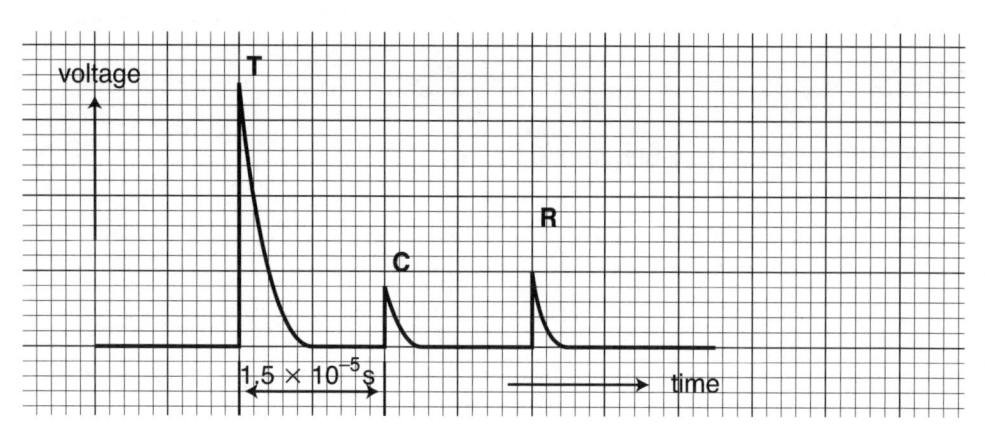

1 Explain why there is an extra pulse **C**.

....

.... **[1]**

2 Use the information given by the oscilloscope trace to calculate the speed of the ultrasound in the metal.

speed = ... m/s **[3]**

3 The frequency of the ultrasound is 3.6×10^8 Hz. Calculate the wavelength of the ultrasound.

wavelength = ... m **[3]**

(Total 10 marks)

Letts

4 **(a)** A rubber balloon acquires a positive charge when it is rubbed by a piece of cloth.

    (i) In terms of electrons, explain how the balloon acquires a positive charge.

    ....

    .... **[2]**

    (ii) The diagram below shows two positively charged balloons which are placed alongside each other.

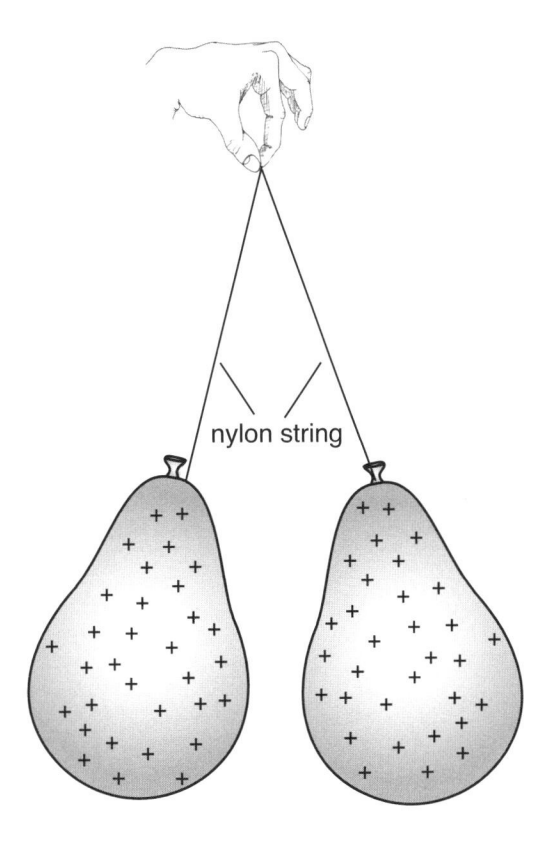

nylon string

    On the diagram above, using arrows show the size and direction of the electric force experienced by each of the balloons. **[2]**

Leave blank

(b) An electric kettle has 350 g of water at 20 °C. The kettle is switched on. The current in the heating element of an electric kettle operating at 230 V is 7.2 A.

Calculate

(i) the resistance of the heating element of the heater,

resistance = ... Ω **[3]**

(ii) the electrical power of the kettle,

power = ... W **[3]**

(iii) how long it takes for the water to reach its boiling point (100 °C). It takes 4200 J of heat to change the temperature of 1 kg of water by 1 °C.

time = ... s **[4]**

(iv) the charge flowing through the kettle.

charge = ... unit: .... **[4]**

(Total 18 marks)

© Letts Educational 2003

5 (a) Explain what is meant by the *half-life* of a radioactive substance.

Leave blank

....

.... **[2]**

(b) The table below shows the half-lives of three emitters of β particles.

nucleus	half-life
strontium-90 ($_{38}^{90}$ Sr)	28 years
lead-214 ($_{82}^{214}$ Pb)	27 minutes
carbon-14 ($_{6}^{14}$ C)	5700 years

(i) Which part of the atom does the β particle come from?

.... **[1]**

(ii) State two properties of β particles.

....

.... **[2]**

(iii) β particles are very dangerous to human beings. Explain why they are dangerous and state how you can minimise exposure from such a radiation.

....

....

.... **[2]**

11

[turn over

Letts

(iv) Samples of the three nuclides have the same number of atoms. Which sample has the highest activity? Explain your answer.

....

....

.... **[2]**

(v) What fraction of the nuclei of lead-214 has **decayed** after a time of 81 minutes?

fraction = ... **[3]**

(Total 12 marks)

6 **(a)** Explain why metals are good conductors of heat.

....

....

....

....

.... **[3+1]**

(b) The gas atoms around us are constantly colliding with each other. The diagram below shows a helium atom making a head-on collision with a stationary carbon atom.

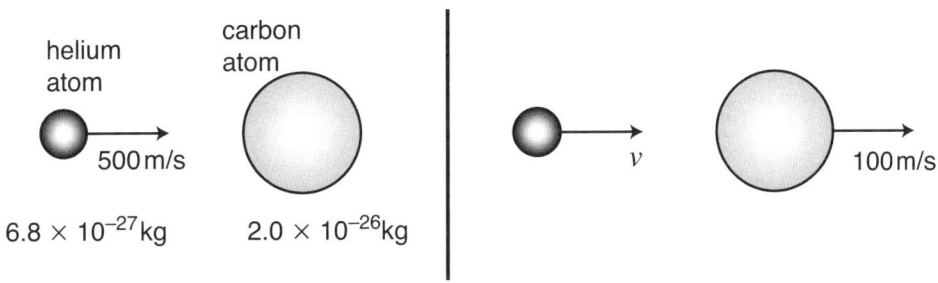

(i) Calculate the initial momentum of the helium atom.

momentum = ... kg m/s **[3]**

(ii) Use the information provided on the diagram above to determine the velocity of the helium atom after its collision with the carbon atom.

velocity = ... m/s **[4]**

(Total 11 marks)

7 **(a)** All logic gates use digital signals. With the aid of a sketch, explain what is meant by a digital signal.

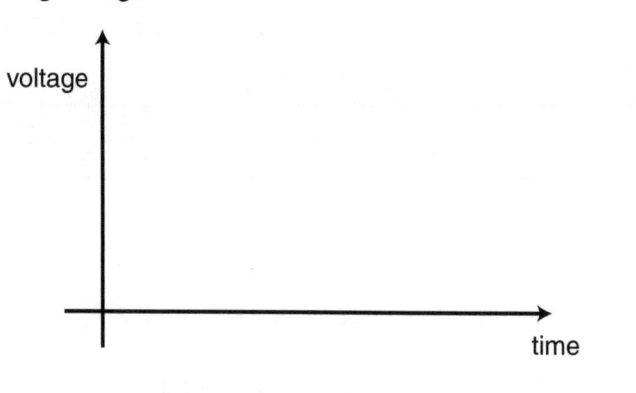

....

.... **[2]**

(b) The diagram below shows a logic gate.

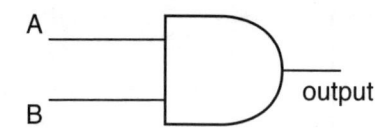

(i) Name the logic gate shown above.

.... **[1]**

(ii) Complete the truth table for the logic gate. **[1]**

A	B	output
0	0	
0	1	
1	0	
1	1	

(c) The diagram below shows a potential divider circuit consisting of a light-dependent resistor (LDR).

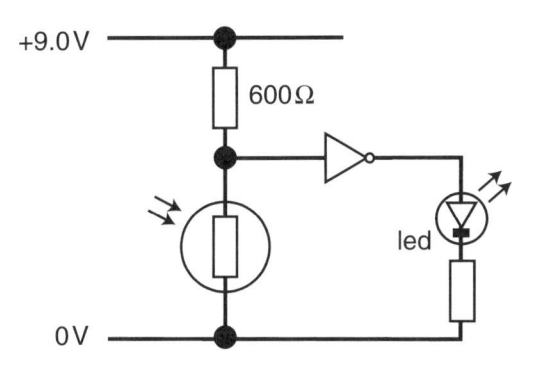

(i) State how the resistance of the LDR is affected by the intensity of light.

....

.... **[1]**

(ii) In sunlight, the resistance of the LDR is 1200Ω. Calculate the voltage across the LDR.

voltage = ... V   **[3]**

(iii) State whether or not the light-emitting diode (led) is lit.

.... **[1]**

(Total 9 marks)

BLANKPAGE

© Letts Educational 2003

Letts Examining Group

General Certificate of Secondary Education

Physics
Higher Tier

Mark scheme
and
Examiner's report

Question Answer	Mark	Question Answer	Mark

1 a

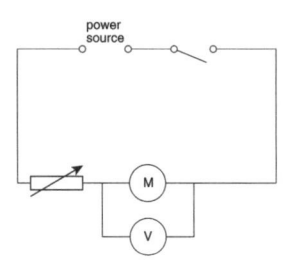

correct symbol for variable resistor	**1**
in series	**1**
voltmeter in parallel with motor	**1**

b increasing resistance **1**
reduces current, reduces force, slows motor **1**
(or decreasing resistance speeds motor up)

2 a equal (in size) **1**
opposite (in direction) **1**

b i weight = 90 × 10 **1**
= 900 N **1** (must have unit)
ii 90 kg **1** (must have unit)
iii weight = 90 × 1.6 **1**
= 144 N **1**

Examiner's Tip
This question tests your understanding of the difference between mass and weight. It is important to know that weight is a force, measured in N, acting on a mass. Mass is measured in kg.

3 a Longitudinal waves: the vibration is in the same plane as the wave motion. **1**
Transverse waves: the vibration is at right angles to the wave motion. **1**

b Sound waves make the air vibrate. **1**
This makes the candle flicker. **1**

Examiner's Tip
When faced with a question about work that you may think you have not have been taught, e.g. about loudspeakers and candles, remember that the examiner is testing your understanding of basic physics. So try to decide which bits of physics the examiner is testing!

c i Ultrasound has a much higher frequency than audible sound. **1**

ii 2 × distance = speed × time
= 1500 × $\frac{0.15}{2}$ **1**
= 112.5 m **2**
If 0.15 s is used for time instead of $\frac{0.15}{2}$, these two marks are still scored. 1 mark for unit.

Examiner's Tip
Remember that the sound has to travel to the shoal and back, so the distance travelled is twice the distance to the shoal.

4 a

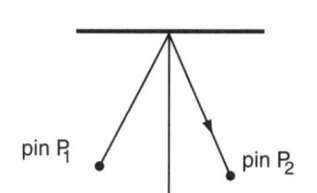

correct position of X	**1**
correct incident ray	**1**
normal with i = r	**1**
arrow on reflected ray	**1**

Examiner's Tip
The best students draw ray diagrams with confidence and care.

b rays of light **appear** to have come from the image **1**

c *1 mark for each correctly drawn wave* **3**

i

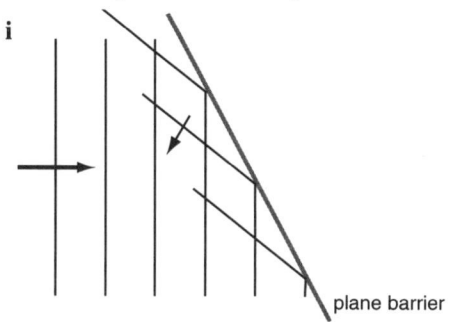

plane barrier

ii

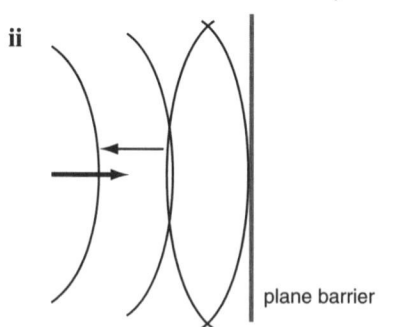

plane barrier

iii

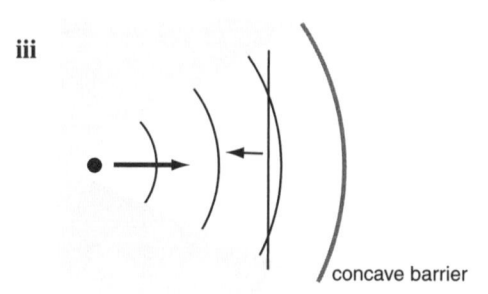

concave barrier

© Letts Educational 2003

2

5 a Microwaves cause heating when absorbed by water in the tea. — **1**

The material of the mug does not contain water so is not heated by microwaves. — **1**

Infrared causes heating by being absorbed by any object. — **1**

b speed = frequency × wavelength

f = speed/wavelength

f = 3 × 10^8/1500 — **1**

= 200 000 Hz — **2** *(1 for unit)*

c i

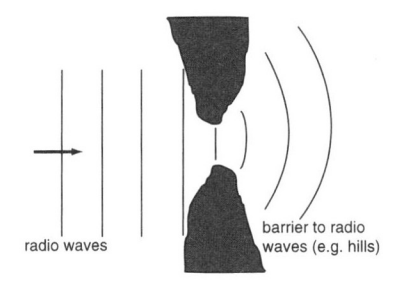

radio waves

barrier to radio waves (e.g. hills)

good/adequate completion of diagram — **2/1**

ii People 'living in the shadow' of the hills can receive the signals. — **1**

The above made very clear by showing on diagram or description. — **1**

Examiner's tip

The best students will be confident in their knowledge and understanding of electromagnetic waves and the use of technical terms like 'diffraction'.

6 a i helium nucleus or 2p and 2n — **1**

+2   2 × charge on electron (positive) — **1**

electron — **1**

−1   charge on electron (negative) — **1**

ii electromagnetic wave — **1**

b i beta — **1**

ii It passes through the brown paper — **1**

but not the aluminium. — **1**

iii use Geiger tube with all sources out of the way — **1**

measure count rate for a period of time (at least 2 mins) — **1**

7 a Negatively charged — **1**

electrons from atoms on the surface of the cloth — **1**

are transferred to the polythene surface. — **1**

b All hairs become positively charged — **1**

and repel each other. — **1**

8 a i zero — **1**

ii There was greater air resistance (caused by the bicycles on the roof rack) — **1**

so more energy was required to overcome this. — **1**

b i *Triangle shown* — **1**

gradient = $^{10}/_8$ = 1.25 — **1**

acceleration = 1.25 m/s^2 — **2** *(1 for unit)*

ii F = ma

= 900 × 1.25 — **1**

= 1125 N — **2** *(1 for unit)*

9 a

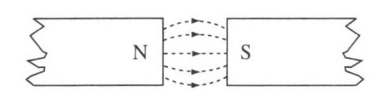

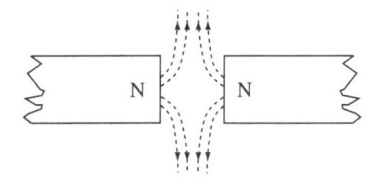

1 mark for shape of field, 1 mark for correct direction shown (does not need to be on every line but must be convincing and consistent): 2 marks for each diagram = total. — **4**

b 1 increased current — **1**

2 more turns — **1**

3 iron core — **1**

c i electromagnet magnetised — **1**

attracts iron armature — **1**

closes contact — **1**

completes circuit to supply current to motor — **1**

+ 1 mark for correct sequence — **1**

ii the current is much greater — **1**

thinner wires would burn out — **1**

10a i

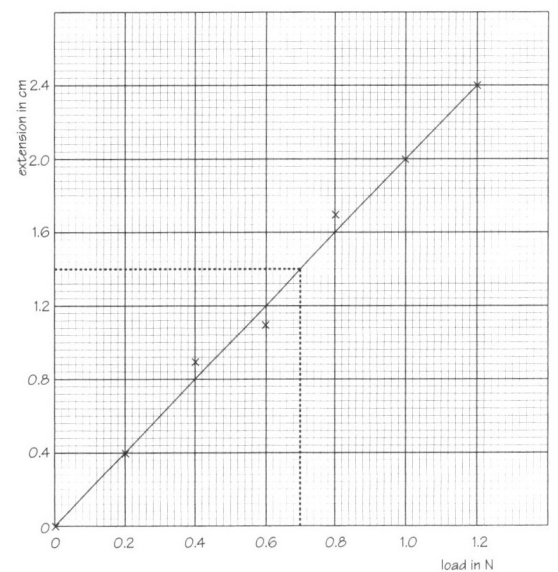

correct axes, labelled — **1**

Question	Answer	Mark
	6/7 correct plots $\quad$ **2** *(4/5 correct = 1)*	
	good line judgement	**1**

Examiner's Tip
If you judge that the points show a straight line, draw the line with a rule. The line you draw should be no thicker than the thick lines on the graph paper.

ii	It obeys Hooke's Law, or extension proportional to load.	**1**
iii	1.4 cm *correct reading from graph*	**1**
	clear on graph	**1**
iv	WD = force × distance	**1**
	$= 0.6 \times 2.4 \times 10^{-2}$	**1**
	= 14.4 mJ $\quad$ **1** *(deduct a mark if cm used)*	
b	$0.004 = 0.5 \times 0.004 \times v^2$	**1**
	$v^2 = 2$	**1**
	$v = 1.4$ m/s	**1**

Question	Answer	Mark
11a	so that they can be switched on independently	**1**
b i	P = VI	
	$= 240 \times 4$	**1**
	= 960 W	**1**
	unit correct	**1**
ii	Q = It	
	$= 4 \times 5 \times 60$	**1**
	= 1200 C	**1**
	unit correct	**1**
iii	240	**1**
c	Direct current is always in the same direction;	**1**
	alternating current changes direction.	**1**

Examiner's Tip
Notice how important it is to understand the difference between series and parallel circuits and the relationships between voltage, current, charge, energy and power.

Answers: GCSE Physics exam 1 paper 2

Question	Answer	Mark
1 a i	zero	**1**

Examiner's Tip
Before the cork is shot out, its velocity is 0, so the momentum is also 0. Notice that you are not asked to calculate and there is no space given for working out.

ii	momentum $= 0.5 \times 3$	**1**
	= 1.5 gm/s *(answer = 1, unit = 1)*	**2**
iii	the 'pop' gun moves	**1**
	'backwards' or in the opposite direction	
	to the cork	**1**
b	pV = constant or $p_2 \times V_2 = p_1 \times V_1$	**1**
	$p_2 = \dfrac{1.1 \times 60}{30}$	**1**
	= 2.2 Pa	**1**
c	The pressure of the air in the tube is due to the bombardment of the walls by fast moving air particles.	**1**
	When the volume is reduced the particles hit the walls more frequently	**1**
	so the pressure is increased.	**1**
2 i	The weight of an object acts from its centre of mass.	**1**
ii	moment $= 50 \times 25$	**1**

Question	Answer	Mark
	= 1250 Ncm (1.25 Nm)	**1**
iii	$50 \times 25 = W \times 10$	
	$W = \dfrac{50 \times 25}{10}$	**1**
	W = 125 N	
	answer = 1, unit = 1	**2**
iv	(to the left) away from the pivot	**1**
3 a	Half-life is the time for the number of undecayed nuclei to halve.	**2**

Examiner's Tip
Many candidates think that the half-life is half of the sample's life. This is not the case.

b i	C	**1**
	long half-life	**1**
	so that smoke detector works for many years	**1**
	low penetrating power	**1**
	so that smoke absorbs radiation	**1**
	+ 1 mark for an answer in sentences with correct punctuation and grammar.	**1**
ii	F	**1**
	long half-life	**1**
	so that gauge works for many years	**1**
	sufficiently (medium) penetrating	**1**
	for some radiation to be absorbed by the foil	**1**

Letts

4 a i

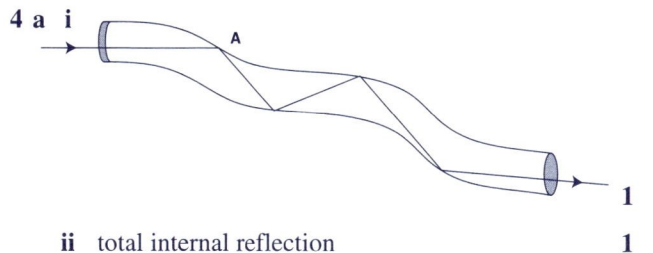

1

ii total internal reflection   1

b i

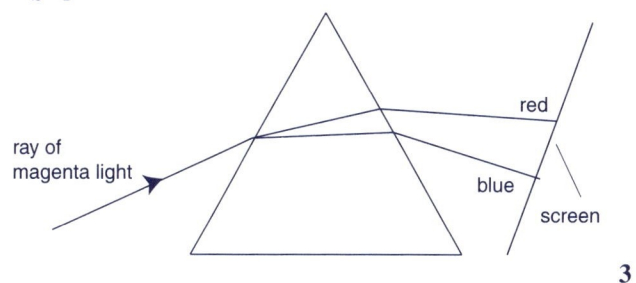

3

ii dispersion   1

iii 2   1

5 a speed has size (magnitude) only   1
velocity has size (magnitude) and direction   1
(or velocity is speed in a given direction)

b i acceleration = $\dfrac{\text{increase in velocity}}{\text{time}}$ 1

$$= \frac{5-3}{12}$$
$$= 1.67\,\text{m/s}^2$$
answer 1, unit 1   2

ii The cyclist was decelerating or slowing down.   2

Examiner's Tip
The best students are confident with the technical terms of physics.

---

6 i

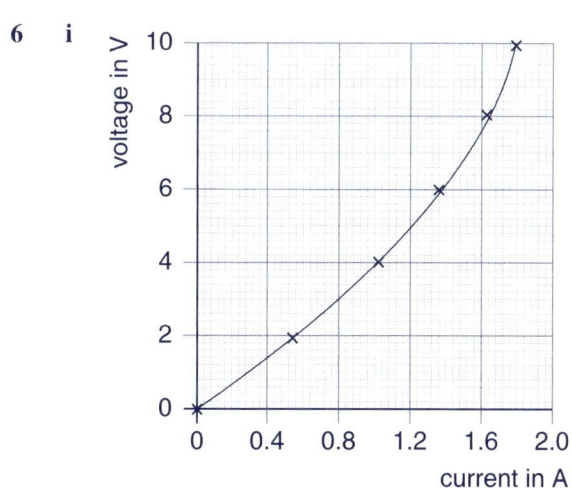

correct axes, labelled   1
plot at 0,0 included   1
other plots correct   1
well judged curve   1

---

ii the resistance increases   1

iii the colour changes/it gives out light   1

iv

1

7 i

particles close together   1
no pattern   1

ii

particles not touching   1
*distances between particles clearly much larger
than the size of the particles*   1

Examiner's Tip
It is important to realise that the particles in a liquid are closely packed together (which is why a liquid is incompressible) but they are not arranged in a regular pattern, whereas in a solid the particles are closely packed together and in a regular pattern.

---

iii solids: the particles vibrate (about fixed positions) 1
liquids: the particles are free to move over and round each other   1
gases: the particles move very quickly in all directions   1

Question	Answer	Mark

1 a i Kinetic energy of the wind to electrical energy. — **1**

ii Generates energy only when it is windy or noise pollution. — **1**

iii Renewable. — **1**

iv efficiency = useful energy ÷ input energy — **1**
$0.74 = 50 \div$ input power — **1**
input power $= 50 \div 0.74 = 68\,\text{kW}$ — **1**

Examiner's tip:
The equation for efficiency is normally written as the ratio of energies, but remember power is energy per unit time. Hence the same equation applies to ratio of powers as well. In order to get the correct answer, it is sensible to set up the equation and then solve it in terms of the input power. There is no need to convert the kilowatts into watts because the final answer should be in kilowatts.

b i kinetic energy $= \dfrac{1}{2} \times$ mass $\times$ velocity^2

$(E = \dfrac{1}{2}\text{mv}^2)$ — **1**

kinetic energy $= \dfrac{1}{2} \times 600 \times 14^2$ — **1**

kinetic energy $= 58\,800 \approx 5.9 \times 10^4$ — **1**
unit: joule (J) — **1**

Examiner's tip:
You should learn the units for all physical quantities. However, if you cannot remember the unit for kinetic energy, you can work it out from the equation: $E = \frac{1}{2}\text{mv}^2$. The unit will be: kg $\times$ (m/s)$^2 \rightarrow$ kg m^2/s^2
This looks unfamiliar and complicated, but it is correct and the examiner will have to give you full credit. But it is easier to recall that the unit for energy is the joule.

ii 58800 J — **1**

Examiner's tip:
The energy is conserved. Since there are no losses due to friction (heat losses), all the kinetic energy is transferred into gravitational potential energy.

iii gravitational potential energy = mass $\times$ gravitational field strength $\times$ height
(PE = mgh) — **1**
$58800 = 600 \times 10 \times H$ — **1**

$H = \dfrac{58800}{6000} = 9.8\,\text{m}$ — **1**

Examiner's tip:
As an A grade candidate, you must know how to rearrange an equation.

2 a There is a reaction from the board vertically upwards. — **1**

Examiner's tip:
The swimmer is standing still; therefore the net force must be zero. You already know that the weight acts downwards hence the reaction must be equal but opposite to the weight.

b weight = mass $\times$ gravitational field strength
($W = \text{mg}$) — **1**
weight $= 62 \times 10 = 620\,\text{N}$ — **1**

c i Clockwise moment — **1**

ii The force F is greater than the weight because the distance of the force F from the pivot **X** is smaller. — **1** **1**

Examiner's tip:
moment = force $\times$ distance from pivot
The clockwise moment due to the weight is equal to the anticlockwise moment due to the force F. Since the distance of the force F from the pivot is smaller than the distance of the weight from the pivot, the force F must be larger than the weight.

d i The swimmer has a <u>constant</u> acceleration up to 1.5 s. — **1+1**
After 1.5 s the swimmer is in the water and decelerates. — **1**
Any further detail. — **1**
(E.g.: The deceleration is not constant or the swimmer slows down because of resistance from the water)
+ 1 mark for an answer in sentences with correct punctuation and grammar. — **1**

Examiner's tip:
There are two marks for the first statement.
For writing down '*the swimmer accelerates*' you will only get one mark.

ii acceleration = rate of change of velocity
or acceleration = gradient of graph — **1**
$a = \dfrac{10-0}{1.0}$ — **1**
$a = 10\ \text{m/s}^2$ — **1**

Examiner's tip:
Always show your working. If you do the question by calculating the gradient from the velocity against time graph, then show the 'triangle used' on the actual graph. Examiners would expect working here since the acceleration for free fall is known by most candidates.

3 a The oscillations are at right angles to the wave direction. **1**

b i The speed of light decreases as it enters the plastic. **1**

ii

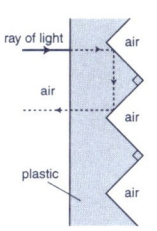

Ray internally reflected at the first interface. **1**
Ray internally reflected at second surface and emerging parallel to the original ray. **1**

c The sound is diffracted at the open door. **1**
Light is not diffracted at the door. **1**
Any further detail. **1**
(E.g.: The sound from the piano has a wavelength similar to the width of the door or the wavelength of light is too small to be diffracted)

Examiner's tip:
The last mark is definitely for an A grade candidate.

d i speed = frequency × wavelength $(v = f\lambda)$ **1**
$$\lambda = \frac{v}{f} = \frac{340}{220}$$ **1**
wavelength = 1.54 m ≈ 1.5 m **1**

ii The sound from both loudspeakers interferes destructively. **2**
(Allow one mark for writing '*waves cancel each other*' or '*peaks and troughs combine*'.)

4 a

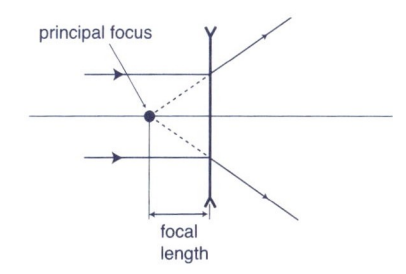

Both rays shown to diverge at the lens. **1**
Correct distance marked for the focal length. **1**

b i An image that cannot be formed on a screen. **1**

Examiner's tip:
Resist writing '*virtual image is not real*', unless you aim to explain what is meant by a real image.

ii

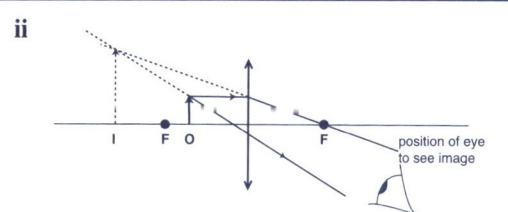

Correct path for one ray. **1**
Correct path for second ray. **1**
Image located correctly and labelled **I**. **1**

iii The image is magnified **1**
and upright. **1**

5 a i Radiation **1**

ii Conduction **1**

iii number of Units = power in kW × time in hours **1**

number of Units = $1.2 \times \frac{10}{60}$ = 0.2 kW h **1**

total cost = number of Units × cost per Unit
total cost = 0.2 × 7.2 = 1.4 pence **1**

b i Parallel **1**

ii current = voltage ÷ resistance $(I = \frac{V}{R})$ **1**

current = $\frac{12}{30}$ **1**

current = 0.4 A **1**

Examiner's tip:
In a parallel circuit, the voltage across each component is the same.

iii power = voltage × current $(P = VI)$ **1**
power = 12 × 0.4 = 4.8 W **1**
energy = power × time
energy = 4.8 × 60 = 288 J ≈ 290 J **1**

Examiner's tip:
In order to calculate the energy released by the resistor, the time must be converted into seconds. Failure to do so will lose you at least one mark.

Letts

6 a The current in the coil **A** creates a magnetic field in the soft-iron rod. **1**
This magnetic field is changing when the switch is closed. **1**
This changing magnetic field links the coil **B** and hence a voltage is produced in coil **B**. **1**
Eventually, the constant current produces a constant magnetic field. There is no change in the magnetic field, hence there is no voltage produced in coil **B**. **1**
+ 1 mark for an answer in sentences with correct punctuation and grammar. **1**

Examiner's tip:
This is a tough question. It is testing whether or not you understand how a 'transformer' works.
Learn this section well.

b i $\dfrac{\text{voltage across primary}}{\text{voltage across secondary}} = \dfrac{\text{turns on primary}}{\text{turns on secondary}}$ **1**
$230 \div 3.8 = 5200 \div \text{turns on secondary}$ **1**
$\text{turns on secondary} = \dfrac{5200}{230} \times 3.8 = 86$ **1**

Examiner's tip:
Candidates often make a mess of rearranging this equation. You know that it is a step-down transformer. Hence the number of turns on the secondary will be smaller than that on the primary coil by a factor of
$\dfrac{230}{3.8} = 60.5$
Therefore, the number of turns on the secondary coil
must be $\dfrac{5200}{60.5} = 86$

ii input power = output power **1**
$230 \times 0.012 = 3.8 \times \text{current}$ **1**
$\text{current} = \dfrac{230 \times 0.012}{3.8} = 0.73\,\text{A}$ **1**

Examiner's tip:
In a step-down transformer, the voltage is stepped down but the current increases. The current increases by the same factor as the ratio of the number of turns.
Hence current in the secondary coil
$= 12\,\text{mA} \times 60.5 = 730\,\text{mA} = 0.73\,\text{A}$

c High voltage means smaller current therefore less energy is lost in the overhead cables as heat. **1**

7 a An element, the nucleus of which has the same number of protons **1**
but different number of neutrons. **1**

b Background radiation is radiation produced by external sources. **1**
Rocks (granite) or outer space or nuclear fall out or medical sources. **1**

c i protons $= 6$ **1**
neutrons $(= 14 - 6) = 8$ **1**

ii $90 \div 22.5 = 4$ **1**
number of half-lives $= 2$ **1**
age $= 5600 \times 2 = 11200$ years **1**

Examiner's tip:
The question is definitely something to do with half-lives. You can start with 90 counts per hour and keep dividing the answer by 2 and stop when you reach 22.5 count per hour. The number of times you have to do this will give you the number of halves-lives. Therefore:
$90 \div 2 = 45 \;\rightarrow\; 45 \div 2 = 22.5$
one half-life   another half-life

8 a i Sun emits light whereas a planet reflects light. **1**

ii Gravitational force. **1**

b Man-made objects that orbit the Earth. **1**
Communications/weather monitoring/spying. **1**

c i The direction changes, therefore the velocity changes. **1**

Examiner's tip:
Velocity has both size and direction. It is a vector quantity.

ii1 Acceleration that is always directed towards a fixed point
(in this case, the centre of the Earth) **1**

ii2 radius $= 6\,700\,000\,\text{m}$ and speed $= 7700\,\text{m/s}$ **1**
$a = \dfrac{v^2}{r} = \dfrac{7700^2}{6\,700\,000}$ **1**
$a = 8.8\,\text{m/s}^2$ **1**

Examiner's tip:
Before substituting your numbers into the equation, you must first convert the radius into metres and the speed into metres per second.

Answers: GCSE Physics exam 2 paper 2

Question	Answer	Mark

1 a i Position **C**. 1

Examiner's tip:
It is best not to guess here. The air resistance depends on the speed of the object. The drag force increases as the speed of the object increases. The question wants you to work out in which position the speed of the child is a maximum. The speed of the child is a maximum at position **C**.

ii *Any 2 from:*
Gravitational potential energy to kinetic energy from **A** to **C**.
Kinetic energy to gravitational potential energy from **C** to **D**.
Kinetic energy (or potential energy) to heat as the oscillations die out. **1+1**
+ *1 mark for an answer in sentences with correct punctuation and grammar.* 1

iii Metal is a good conductor of heat. 1
Heat flows from the hands to the chains, so the hands feel cold. 1

b i 15% 1

ii Loss through the flooring can be reduced by having a thicker carpet or an underlay. 1

iii Loss of heat by conduction is reduced 1
because air is a poor conductor of heat. 1

iv The shiny aluminium reflects radiation back into the house. 1

2 a i The thinking distance is the distance travelled by the car whilst the driver is reacting. 1
The braking distance is the distance the car travels while the brakes are applied and the car stops. 1

Examiner's tip:
Candidates often forget to mention the important word **distance** in both definitions. It would be wrong to state that *'Thinking distance is how long it takes for the driver to react'* because the *'how long'* implies time. Always be very careful with definitions.

ii Any one factor from: road surface, efficiency of brakes, speed of car, mass of car and tyre conditions. 1
Any further detail on one of the factors.
(E.g: The greater the speed of the car, the greater is the braking distance or worn out tyres increase the braking distance of the car because of poor grip between the tyre and the road.) 1

b i acceleration = rate of change of velocity 1
$$a = \frac{0-20}{4.2}$$ 1
$a = -4.76$ m/s$^2 \approx -4.8$ m/s^2 1

Examiner's tip:
The velocity of the car decreases, therefore it is vital to have a **negative** sign for the deceleration. In order to avoid errors in examinations, write down the change in velocity as: 'final velocity – initial velocity'.

ii force = mass × acceleration ($F = ma$) 1
$F = 940 \times -4.76$ 1
$F = 4474$ N ≈ 4500 N (Ignore the sign) 1

iii area under a velocity against time graph = distance 1
distance $= \frac{1}{2} \times 20 \times 4.2$ 1
distance $= 42$ m 1

iv work done = force × distance moved in direction of force 1
work done $= 4474 \times 42$ 1
work done $= 1.9 \times 10^5$ 1
unit: joule (J) or Newton metre (Nm) 1

Examiner's tip:
It is always easier to write your answer in standard form, especially when the numbers are large (or small).

v The braking force is the same. Therefore the deceleration remains constant.
The time taken to stop is doubled. 1
braking distance $= \frac{1}{2} \times$ initial velocity $\times$ time
Since the time is doubled and the initial velocity is also doubled, the braking distance increases by a factor of four. 1

Examiner's tip:
This is a tough question. However, there are always clues left behind by examiners. This question is an extension of what you have already done in (iii). It is important to appreciate that the braking distance is directly proportional to the product of the initial velocity and the time taken for the car to stop. A common mistake would be to assume the braking distance is directly proportional to the initial velocity of the car and therefore end up with a wrong statement *'the distance doubles because the velocity doubles'*.

© Letts Educational 2003

3 a There is a vacuum between the Earth
and the Moon. **1**

Examiner's tip:
It is very easy to use the wrong word and fail to secure a
mark. The key word in the marking scheme is '*vacuum*'.
You may use alternatives like: '*There is no air between the
Moon and the Earth*' or '*Sound needs air to travel*'.
Examiners tend not to like the use of the word '*space*' to
mean '*vacuum*'.

b A high frequency longitudinal wave (sound)
that humans cannot hear. **1**

c i Some of the wave energy lost. **1**

ii 1 Ultrasound reflected from the crack. **1**

ii 2 distance travelled by the ultrasound
$= 2 \times 2.8 = 5.6\,\text{cm}$ **1**
speed $=$ distance $\div$ time **1**
speed $= \dfrac{0.056}{1.4 \times 10^{-5}} = 4000\,\text{m/s}$ **1**

Examiners tip:
A common error made with a question like this is
determining the total distance travelled by the reflected
wave. The wave is reflected by the crack. The total distance
travelled by the wave is twice the depth of the crack from
the surface of the metal. The answer for the speed is in m/s,
therefore do not forget to convert the distance into metres.

ii 3 speed $=$ frequency $\times$ wavelength $\quad (v = f\lambda)$ **1**
$\lambda = \dfrac{v}{f} = \dfrac{4000}{3.6 \times 10^{8}}$ **1**
wavelength $= 1.1 \times 10^{-5}\,\text{m}$ **1**

Examiner's tip:
To be an A grade candidate, you must be comfortable doing
calculations using standard form. Some candidates have
problems either recalling or rearranging the wave equation.
As an A grade candidate, you cannot afford to do this.

4 a i The rubbing action removes electrons
from the balloon. **1**
Electrons have a negative charge. The balloon
is left with a net positive charge. **1**

ii

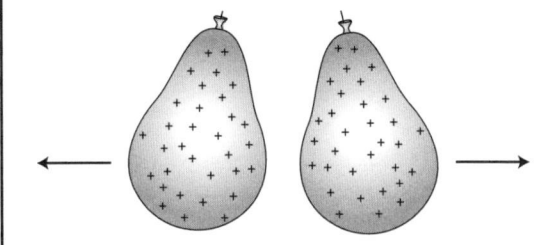

Two arrows in opposite directions. **1**
Equal sized arrows (showing the same force
experienced by each balloon) **1**

b i resistance $=$ voltage $\div$ current $\quad (R = \dfrac{V}{I})$ **1**
$R = \dfrac{230}{7.2}$ **1**
$R = 31.9\,\Omega \approx 32\,\Omega$ **1**

ii power $=$ voltage $\times$ current $\quad (P = VI)$ **1**
$P = 230 \times 7.2$ **1**
$P = 1700\,\text{W}$ **1**

iii energy needed to change the temperature of
water by $1°C = \dfrac{350}{1000} \times 4200$ **1**

energy needed to boil the water
$= \dfrac{350}{1000} \times 4200 \times (100 - 20)$ **1**

energy needed to boil the water
$= 1.18 \times 10^{5}\,\text{J}$ **1**

time $=$ energy $\div$ power $= \dfrac{1.18 \times 10^{5}}{1700}$ **1**

time $= 69\,\text{s}$

iv charge $=$ current $\times$ time $\quad (Q = It)$ **1**
charge $= 7.2 \times 69$ **1**
charge $= 500$ **1**
unit: coulombs or C **1**

Examiner's tip:
You can always work out what the unit for charge from the
equation $Q = It$. The unit for charge is:
$$\text{ampere} \times \text{second} \rightarrow \text{As}$$
If instead of the 'coulomb' you write down either 'ampere
second' or 'As', then the examiner has to give you full
credit for the correct physics.

5 a The half-life of a substance is the average time taken for half **1**
the number active nuclei to decay. **1**

Examiner's tip:
Quite often candidates will know that this definition involves something being halved. You have the choice of stating either '*number of active nuclei*' or the '*activity*'. Reference to the nuclei is important. It would be wrong to state that '*half-life is the time taken for the substance to halve its mass*'.

b i The nucleus. **1**

Examiner's tip:
Do not simply state that 'atoms' release β particles. It is very important to appreciate that radioactivity is to do with the changes taking place within the nuclei. Some candidates wrongly assume that the β particles are the orbiting electrons within the atoms.

ii *Any two from:* **1+1**
They are electrons.
Carry a negative charge.
Can be stopped by a thin sheet of aluminium.

iii They destroy living cells. **1**
Use a thin sheet of aluminium to shield from the source. **1**

iv The lead sample will have the largest activity. **1**
The nuclei decay in a short period of time (compared with the other samples). **1**

v 81 minutes = 3 half-lives **1**

fraction of nuclei left after 81 minutes =

$$\frac{1}{2} \times \frac{1}{2} \times \frac{1}{2} = \frac{1}{8}$$ **1**

fraction of nuclei that have decayed =

$$1 - \frac{1}{8} = \frac{7}{8}$$ **1**

Examiner's tip:
The question requires careful reading because it does not want you to just determine the fraction of nuclei **left** within the sample after 81 minutes. A mathematical way of getting the answer would be to use:

number of nuclei decayed = $1 - (\frac{1}{2})^n$,

where n = number of half-lives.

6 a Metals are good conductors because of free electrons. **1**
These electrons can easily diffuse through the metal. **1**
Electrons gain thermal energy by colliding with the vibrating atoms at the hot end of the metal and transport this energy to the cooler atoms at the cold end. **1**
+ *1 mark for an answer in sentences with correct punctuation and grammar.* **1**

b i momentum = mass × velocity **1**
momentum = $6.8 \times 10^{-27} \times 500$ **1**
momentum = 3.4×10^{-24} kg m/s **1**

Examiner's tip:
Momentum is a vector quantity. It is therefore very important that you use 'velocity' in this equation and not 'speed'. Use of the term speed in the equation will lose you the first mark.

ii initial momentum = final momentum **1**
$3.4 \times 10^{-24} = (6.8 \times 10^{-27} \times v) +$
$(2.0 \times 10^{-26} \times 100)$ **1**
$v = (3.4 \times 10^{-24} - 2.0 \times 10^{-24}) \div$
6.8×10^{-27} **1**
$v = 206$ m/s **1**

Examiner's tip:
The key to doing this type of question is to set up an equation using the law of conservation of momentum and then solve for the unknown velocity v.

7 a

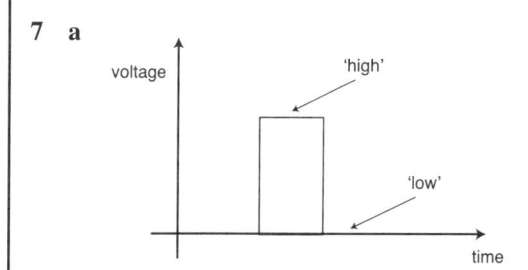

Correct sketch for a digital signal (see the sketch above). **1**
A digital signal has only two values or levels of voltage. **1**

Examiner's tip:
You can also state that a digital signal only has 'high' or 'low' values.

b i AND gate **1**

ii

A	B	output
0	0	0
0	1	0
1	0	0
1	1	1

1

c i

The resistance of the LDR decreases as the intensity of light increases. **1**

ii total resistance $= 1200 + 600 = 1800\,\Omega$ **1**

current $= \dfrac{9.0}{1800} = 0.005\,\text{A}$ **1**

voltage $= IR = 0.005 \times 1200 = 6.0\,\text{V}$ **1**

Examiner's tip:
The current in a series circuit is the same. Use this idea to calculate the voltage across the LDR and do not forget to use the resistance of the LDR and not the resistor. Using $600\,\Omega$ would give the wrong answer. If however, you show all your working, then you would only lose the final mark.

iii The led is not lit. **1**

Examiner's tip:
The NOT gate inverts the input. The input voltage is $6.0\,\text{V}$, which is closer to the $9.0\,\text{V}$ than the $0\,\text{V}$. Therefore the input to the gate is 'high'. The NOT gate changes this high input into a low output. Consequently, the light-emitting diode is not lit.

WHAT IS IN THE EXAM?

The first examinations of new specifications (or syllabuses) were held in 2003.

For all Physics GCSE courses candidates had to complete two papers – one in the Core material (common with the Physics in Double Award Science) and one which included extra extension material added to make up the Physics specification. Each Awarding Body (or Examining Board) added different extension material. The extension material is not more difficult but gives each specification its identity.

The marks in these two papers added together make up eighty per cent of the marks, with twenty per cent coming from the Coursework (Sc1). The grade is then calculated from the total mark achieved.

From 1998 questions on the Higher tier have tested the whole specification. Some statements are identified in the specification as only to be examined on Higher tier but the examinations test the whole specification not just these statements. From 2003, questions could not be set on material you covered in Key Stage 3.

Most people think that GCSE examinations concentrate on testing the knowledge you have learnt. This was certainly the case twenty or thirty years ago. Then examinations concentrated on testing recall of knowledge and there was much more factual information on the syllabus. This is why people keep saying that the examinations are getting easier. You now have to recall fewer facts but you are required to do much more with the knowledge you have.

Ignoring the Coursework, of the remaining eighty per cent of the marks, just sixty per cent is allocated to knowledge and understanding. Of this, one third is allocated to recall of facts and the remainder to showing an understanding of the factual knowledge. Being able to recall Ohm's Law is knowledge, but being asked to describe how resistance changes when current and voltage change is testing understanding rather than recall.

The final fifteen per cent is allocated to higher level skills of interpretation, evaluation, etc. This could involve using your knowledge from different parts of the specification and applying it to a different and new situation. It is here that candidates find the most difficulty.

Examination papers have to have questions requiring longer answers. These are called Continuous and Extended Writing. For Higher tier about 15 marks have to require answers of two sentences and 10 marks have to require longer answers. It is common on candidates' papers to see that performance on these questions is disappointing compared with the rest of the papers. Where candidates often go wrong is to fail to give complete answers and to get the answer out of the correct order.

If you are having difficulty in Physics and your aim realistically is to achieve a Grade C, you are more likely to achieve this by taking the Foundation tier papers.

NEW FEATURES IN 2003

Quality of Written Communication (QWC)

About four marks on each paper have to be allocated to QWC. On these questions – and you will be told which questions they are – marks can be awarded by the examiner for your ability to either:

- write in sentences
- use correct spelling, punctuation and grammar
- use correct scientific terms.

Many candidates write answers in bullet points rather than sentences. You should make sure you use sentences in any of these questions.

Ideas and Evidence questions

Five per cent of the questions on the paper will be based upon Ideas and Evidence. You will probably not know which questions they are. You are expected to use the information to give your opinions but you should always support your opinions with scientific information either from the paper or from your knowledge.

Ideas and Evidence can be divided into four sections.

1. How scientific ideas are presented, evaluated and disseminated.

2. How scientific controversies can arise from different ways of interpreting empirical evidence.

3. Ways scientific work may be affected by the context in which it takes place.

4. Ways to consider power and limitations of Science in addressing industrial, social and environmental questions.

HOW TO IMPROVE YOUR GRADE

If you are taking GCSE Physics, your grade will depend upon your performance in the two written papers and Coursework.

Frequently candidates who do well on the Core Physics paper do less well on the Extension paper. There can be many reasons for this. Possibly less time has been allocated for this extension material both in school or college but also in revision.

An examination cannot cover everything in the specification but it does try to sample it all. Do not miss out sections in your revision. Concentrate on the whole specification and pay special attention to learning definitions and the correct use of scientific terminology.

WHAT IS NEEDED FOR A GRADE A?

Contrary to many people's beliefs, grade A is not determined each year by awarding the grade to a fixed percentage of candidates or by awarding a grade A to those candidates who achieve a fixed mark. It is done by inspection of the papers and awarding grade A to those candidates who meet the criteria that have been agreed nationally for grade A.

A grade A candidate should be able to:

1. use detailed knowledge and understanding to devise a strategy for a task;

2. identify key factors in the task and control conditions;

3. make predictions;

4. present data appropriately and use knowledge from different sources;

5. recognise and explain anomalous results;

6. draw appropriate graphs choosing suitable axes;

7. use scientific knowledge and understanding to draw conclusions;

8. identify shortcomings in evidence;

9. use a range of apparatus with the correct precision and skill:

10. make precise measurements and systematic observations;

11. select which observations and measurements are relevant;

12. recall information from all areas of the specification;

13. use detailed scientific knowledge and understanding in a range of applications;

14. detect patterns and draw conclusions when information comes from different sources;

15. draw information together and communicate knowledge effectively;

16. use scientific or mathematical conventions to support arguments;

17. use a wide range of scientific and technical vocabulary.

Some of these criteria can be met in Coursework (Scl) but most can also be demonstrated in written papers.

WHAT EXTRA IS REQUIRED FOR AN A* GRADE?

Having established what is required to be awarded a Grade A you might be interested to know what is required for an A* grade. There are at present no A* criteria.

When the Awarding committee is awarding grades it is asked to fix marks for Grade A and Grade C. This is done paper by paper. Suppose on a particular paper the Grade A mark was fixed at 70 and the grade C mark at 50. (These numbers have been chosen only to keep the arithmetic that follows simple). The Grade B mark is then fixed arithmetically half way between 50 and 70, i.e. 60. The Grade A* is then fixed the same number of marks above A as B is below it. In the example we have used A* would be fixed at 80. At this point it would be customary to look at papers around this mark to confirm that they were worthy of A*. What does this tell you? Grade A* is a very high standard and relatively few are awarded.

As there are no criteria it is not as clear what examiners are looking for as it is at Grade A. The following points may help you.

- Generally as the Grade A* boundary is a high mark, there is no scope for a bad answer to any question on the paper. A grade A* candidate scores well on questions.
- A grade A* candidate uses scientific language routinely and confidently. It is worthwhile working through a glossary of scientific terms or a scientific dictionary to clarify the exact meaning of all terms and then trying to use them correctly.
- A grade A* candidate brings information from different parts of the specification together in an answer.
- Grade A* candidates in Physics have a clear idea of models of atomic structure and the kinetic model for particles in solids, liquids and gases. They can use these ideas of models to explain ideas such as conduction, convection, a current passing through a wire and radioactivity.
- Grade A* candidates in Physics score highly in calculations including ones where more than one formula has to be used. They also give answers to the correct number of significant places and with correct units.
- While many good candidates can substitute numbers into equations and calculate answers, grade A* candidates understand mathematical relationships. For example, Force = mass × acceleration can be used to calculate the acceleration of a heavily loaded lorry for a given force. A grade A* candidate appreciates that the acceleration of a lorry, with a given force is inversely proportional to the mass of the lorry. Grade A* candidates in Physics can transpose equations and appreciate relationships between quantities such as energy and power. They can demonstrate the understanding of conservation of momentum and perform calculations based upon this with confidence.

HOW TO ASSESS YOUR GRADE

The matrix suggests grades that you might have expected to achieve with different scores on these papers. It is an indication only and does not imply that this is the grade you will receive in the real examination.

	EXAM 1	EXAM 2
A*	120–160	155–180
A	110–119	125–154
B	95–109	100–124
C	80–94	75–99
D	70–79	50–74

Letts Educational
The Chiswick Centre
414 Chiswick High Road
London
W4 5TF

Tel: 020 8996 3333
Fax: 020 8742 8390
Email: mail@lettsed.co.uk

Every effort has been made to trace copyright holders and to obtain their permission for the use of copyright material. The authors and publisher will gladly receive any information enabling them to rectify any error or omission in subsequent editions.

First published 2004

Text, design and illustrations © Letts Educational 2004

All our rights reserved. No part of this publication may be reproduced, stored in a retrieval system, or transmitted, in any form or by any means, electronic, photocopying, recording, or otherwise, without the prior permission of Letts Educational.

No UK examination boards have supplied or approved the questions, answers or grading advice given in this pack; the answers provided may not be the only solutions to the questions given. The results you achieve in the exams in this pack are only an indication of what you may achieve in the official exam.

Prepared by *specialist* publishing services, Milton Keynes

British Library Cataloguing in Publication Data

A CIP record for this title is available from the British Library

ISBN 1843153076

Letts Educational Limited is a division of Granada Learning Limited, part of Granada plc.

Printed in the UK